Giant Hogweed Management in the United Kingdom

Olaf Booy and Max Wade

D1341617

First published in 2007 by RPS Group Plc, Willow Mere House, Compass Point Business Park, Stocks Bridge Way, St. Ives, Cambridgeshire, PE27 5JL.

Printed by CLE Print Ltd, Media House, Burrel Road, St. Ives, Huntingdon, Cambridgeshire, PE27 3LE.

A CIP record is available from the British Library in London.

ISBN: 978-0-906269-04-6

Front cover design by Adam Shephard, front and back images Olaf Booy.

Citation

Booy, O. and Wade, M. (2007) Giant Hogweed Management in the United Kingdom. RPS Group Plc.

RPS Group Plc
Willow Mere House
Compass Point Business Park
Stocks Bridge Way
St. Ives
Cambridgeshire
PE27 5JL

Tel: +44 (0)1480 466335
Fax: +44 (0)1480 466911
Email: rpscm@rpsgroup.com

Acknowledgements

The authors were part of the European Union Framework V 'Giant Alien' project 'Giant Hogweed (*Hercaleum mantegazzianum*) a pernicious invasive weed: developing a sustainable strategy for alien invasive plant management in Europe' funded by an EU Framework V project grant no. EVK2-CT-2001-00128, one outcome of which was the web based European 'Giant Hogweed Best Practice Manual', (Chapter 12, number 19 and useful links).

The Giant Alien Project and Best Practice Manual were considerable help in preparing this manual and we are grateful to the other members of the team including Petr Pyšek, Irena Perglová, Jan Pergl, Lenka Moravcová, Matthew Cock, Šárka Jahodová, Charlotte Nielsen, Hans Peter Ravn, Jan Thiele and Ruediger Wittenberg.

Vicky White, Helen Parish and Peter McKeon of RPS assisted in the research and production of this manual. David Holland (invasive species consultant) provided valuable review and comment.

The authors wish to thank the Environment Agency for their support in the production of this manual and for part-funding its production.

Contents

Giant Hogweed on the banks of the River Brent, north London

1. Introduction

Giant Hogweed is the largest herbaceous plant in the UK. It is a true giant, not only in the size of its leaves, stems and flowers, but also in its ability to spread. Since introduction from its home in the mountains of southwest Russia into the botanical gardens of London in the early nineteenth century (Figure 1.1), it has rapidly spread throughout the UK. Initially botanical collectors and gardeners shared seeds and specimens to plant as exotic curiosities, often in the grounds of wealthy and prestigious estates. To this day Giant Hogweed can still be found in the gardens of Buckingham Palace and Kew in London, at Chatsworth House in Derbyshire and on the streams and rivers around Cawdor Castle in Nairn, Scotland. Latterly, as the popularity of the plant grew, seeds were made more widely available and planted in a wider range of gardens, parks and churchyards across the UK.

Able to thrive in the UK climate and away from natural herbivores and diseases that would have kept its population under control in its home range, Giant Hogweed was able to spread rapidly, dominating the habitats in which it grew. Unfortunately, this characteristic caused little concern to botanists of the time. The negative impacts of the plant's behaviour only became recognised towards the end of the twentieth century, long after the plant had spread exponentially into every corner of the United Kingdom.

Today, Giant Hogweed can regularly be found growing in parks, wasteland, farms and along roads, railways and, in particular, river banks, along which it is easily transported. In 1981, legislation was passed to make it illegal to spread Giant Hogweed in the wild and it is now recognised by the UK's environmental protection agencies as the number two most undesirable weed, after Japanese Knotweed. Its impact on native species apart, the concern about this plant has often focussed on its toxic sap, which can cause serious burns and blistering to human skin. Other significant impacts include erosion to riverbanks, increasing flood risk and reducing amenity value or parks and rivers.

The introduction of legislation to control the spread of the plant, and the health and safety implications caused by the plant's damaging sap, explain why Giant Hogweed also has a serious impact on development sites. Just as with Japanese Knotweed, developers often have to control Giant Hogweed before undertaking their work in order to avoid committing an offence. These control measures can be costly and in the worst case can delay development for several months.

Figure 1.1 Early Giant Hogweed specimen

Despite efforts to control its spread over the last decades of the 20th century, Giant Hogweed is still in an invasive mode across much of the UK, a factor that helps to explain the increase in research into its biology, ecology and control. The authors worked on one such project entitled 'Giant Hogweed: developing a sustainable strategy for alien invasive plant management in Europe' with a consortium of European Giant Hogweed experts. From this and their practical experience of site to catchment scale control strategies in the UK, the authors have developed this manual to disseminate some of the information that we consider most relevant to the key people and organisations working on Giant Hogweed in the UK. Other sources of valuable information have also been developed from this project, including a compendium of scientific texts (Chapter 12, number 30).

This manual is provided for landowners, managers, developers, regulators and practioners of Giant Hogweed across England, Scotland, Wales and Northern Ireland. It is designed to provide clear, practical guidance for the development and implementation of Giant Hogweed management at all scales within the UK. Each chapter can be read as a stand-alone section or in conjunction with others, and boxes summarising key data are provided to help managers gain quick access to the information relevant to them. A contents page and index are included for quick reference and further reading, references and useful websites are listed. A glossary is also provided to help clarify terminology.

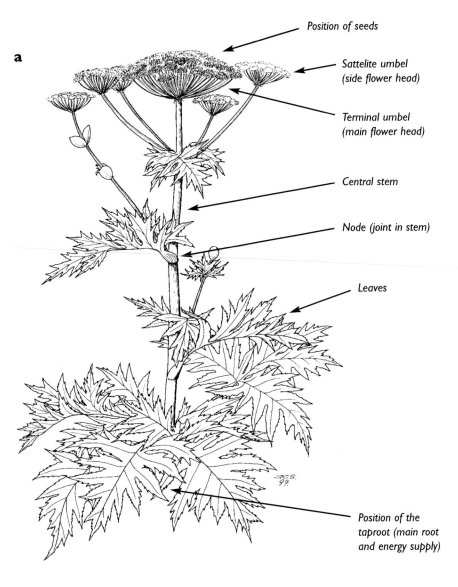

a: Drawing of a Giant Hogweed plant in flower, showing key anatomical features (drawing from J Schou)

b: Inset – taproot (image from Tiley et al. 1996)

Chapter 12, number 38

Position of seeds

Sattelite umbel
(side flower head)

Terminal umbel
(main flower head)

Central stem

Node (joint in stem)

Leaves

Position of the
taproot (main root
and energy supply)

Figure 1.2 The anatomy of a Giant Hogweed plant

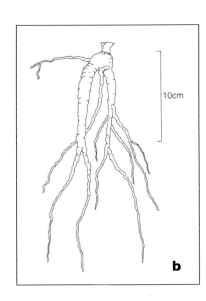

10cm

b

2. Identification

Giant Hogweed is among the tallest herbs in the UK and as such is relatively easily distinguished from other plants when at full height. The most distinctive characteristic is undoubtedly its size; not only height, but also leaves and flower heads. Other key distinctive features include its wide, bristly stem, which is usually completely purple or green with purple blotches and the vast number of seeds that it produces (up to 50,000 per plant). A summary of identification features is provided in Box 2.1.

Giant Hogweed flowers from June to August and releases seed from late August. The seeds germinate from March onwards, quickly developing a taproot (Figure 1.2b). It takes 3 – 4 years for the plant to achieve maturity and in its final year produces the main stem and a large flower head which attains a maximum height of 5m. Stems and leaves die back each winter leaving rigid dead stems in place, which often survive through the following growing season. After flower production the plant dies off completely.

Giant Hogweed is unmistakable when at full height. However there are still some instances in which identification can be difficult. The following section helps to tackle these.

Identifying early growth stages

Plants at the earliest stages of growth can be more difficult to identify as they have not yet reached full size and have no flower heads. During early stages of development Giant Hogweed leaves may be relatively small; however key features can still be used to identify the plants:

- Broad stems and leaves compared to similar native species;
- Green and purple blotchy stem; and
- Dead stems from previous growth of other, possibly parent plants.

Box 2.2 provides illustrations of how to identify Giant Hogweed throughout the year and the life of a plant.

Giant Hogweed growing along the River Usk in Wales (image from Steve Smith)

Box 2.1 Identification features of Giant Hogweed

Stem
- Central stem with side branches
- Up to 5m tall (f)
- 5-10 cm in diameter
- Hollow with nodes (fig 1.2)
- Bristly
- Blotchy or continuous purple (a)

Leaves (b)
- Up to 3m in length
- Bright to lush dark green
- Divided / serrated leaves
- Bristles on underside

Flowers (c)
- Main flower head at top of central stem
- Up to 80 cm across
- White or rarely pinkish
- Umbrella shaped
- Made up of tiny flowers (d)
- Many smaller flower heads on side stems

Seeds (e)
- Up to 50,000 seeds per plant
- Paper thin
- Penny sized
- Oval
- 4 dark stripes (oil ducts) on one side, 2 on the reverse

a: Stem showing purple blotches. b: Leaf. c: Flower head d: Close up of individual flowers (image from Šárka Jahodová). e: Close up of seeds showing different sides. f: Giant Hogweed stem (dead) (image from Olaf Booy and Šárka Jahodová)

Box 2.2 Giant Hogweed throughout the year

Spring	Summer	Winter

*a: Young seedlings. b: 2-3yr old, early spring. c: 2-3yr old, spring. d: 3-4yr old, near flowering. e: Typical flowering plant (image from Steve Smith).
f: Close up of seed head. g: Dead stems in winter.*

Figure 2.1 Dwarf Giant Hogweed plants caused by re-growth after cutting

Figure 2.2 Decaying leaves caused by disturbance (image from Šárka Jahodová)

Figure 2.3 Multiple stems caused by unusual soil conditions

Identifying plants with appearance altered by disturbance

Giant Hogweed plants that have been cut, chemically treated, or otherwise disturbed, may develop abnormal growth forms ranging from a reduction in height (Figure 2.1), to unusually shaped or decaying leaves (Figure 2.2). It may still be possible to identify these plants and the following points should be considered:

- After cutting, Giant Hogweed may re-grow into smaller dwarf plants and can even flower when less than a metre tall (Figure 2.1). Such plants can appear similar to Common Hogweed (see Figure 2.5), however the size of the flower head, leaf and width of stems remain proportionately larger when compared to the native species.

- Dead stems from previous years' growth may provide evidence of Giant Hogweed. Look for tall and wide dead stems with the remains of large dry flower heads still present. Fallen stems may be lying on the ground nearby as well as the remains of standing stems (Box 2.2g).

- Giant Hogweed can react to adverse ground conditions or treatment by producing multiple stems (Figure 2.3). While this may confuse the surveyor, other features such as the size of the plant should allow identification.

Variability in Giant Hogweed leaves

Giant Hogweed leaves often vary considerably in their appearance (Figure 2.4). Surveyors should be aware of the potential for variability in leaf form; however the size of the leaf as well as the other diagnostics of Giant Hogweed (Box 2.1) should make it possible to confidently identify the plant.

Figure 2.4 Variability in Giant Hogweed leaves

Species that could be confused with Giant Hogweed

Giant Hogweed could be confused with other umbelliferous species native to the UK. The most similar is Common Hogweed, a close relative of Giant Hogweed. This and other similar species are described for comparison below:

Common Hogweed *Heracleum sphondylium* (Figure 2.5). Biennial. Stems up to 300cm, ridged and hairy. Leaves 1x pinnate, roundly toothed lobes. Flowers June to September, white/pinkish, white/greenish or white. Outer parts of flower larger than inner parts. Bracts present. Seeds winged disks, 6-10mm (Box 2.3).

Box 2.3 Hogweed species descriptions (adapted from Nielsen *et al.* 2005, drawings from J. Schou)

Plant Species	Stem	Leaf	Flower	Fruit
Giant Hogweed *Heracleum mantegazzianum*	200–5400cm tall. Shaggy (villous) upper stem; lower stem coarsely ridged and more or less hairy. Stem up to 10cm thick at base with purple blotches			
Common Hogweed *Heracleum sphondylium*	60-300cm tall. Lower stem sparsely hairy, upper stem more densely hairy, deeply ridged			

Figure 2.5 Common Hogweed (image from Diane Dobson)

Figure 2.6 a: Wild Parsnip. b: Wild Angelica.
c: Hemlock. d: Cow Parsley.
Images a and d from Diane Dobson. Images
b and c from Biopix.dk

Wild Parsnip *Pastinaca sativa* (Figure 2.6a). Biennial. Stems up to 180cm, furrowed, downy. Leaves 1x pinnate. Flowers July to August, yellow flowers. Bracts absent. Seeds flat and oval with narrow wings, 5-7mm. The sap can also cause phytodermatitis (see Chapter 5).

Wild Angelica *Angelica sylvestris* (Figure 2.6b). Perennial. Stems 200cm or more, purplish with very inflated leaf sheaths. Leaves 2-3x pinnate, with purple nodes. Flowers July to September, white or pinkish flowers. Bracts absent. Seeds oblong with two wings, 4-5mm.

Hemlock *Conium maculatum* (Figure 2.6c). Biennial. Stems up to 200cm, purple blotched. Leaves 2-4x pinnate. Flowers June to July, white flowers. Bracts present. Seeds egg shaped with wings, 2.5-3.5mm. Poisonous (deadly).

Cow Parsley *Anthriscus sylvestris* (Figure 2.6d). Perennial. Stem 60-150cm. Leaves 2-3x pinnate (fern like), feathery. Flowers April to June, creamy/white flowers. Bracts absent. Seeds oblong, 6-9mm.

Hemlock Water-dropwort *Oenanthe crocata*. Perennial. Stems up to 150cm, grooved and slightly kinked. Leaves 1-2x pinnate. Flowers June to July, white flowers. Bracts present. Seeds like elongated red-pepper, 4-5.5mm. Mainly south and west Britain. Poisonous (deadly).

Other Giant Hogweed species

There are at least three different European species of Giant Hogweed, although only *Heracleum mantegazzianum* is considered to be widespread in the United Kingdom. *Heracleum persicum* has been recorded on rare occasions. No taxonomical key has been developed to distinguish between these species, although further guidance can be found in the Giant Alien Best Practice Manual and Pyšek *et al.* 2007 (Chapter 12, numbers 19 and 30).

3. Origins and History

Heracleum mantegazzianum in Europe originated from the Western Caucasus in southwest Russia (Figure 3.1). It was named after an explorer, Paulo Mantegazza, and first described in Italy in 1895 by Stephan Sommier and Emile Levier. However, according to botanical records, the history of this plant's introduction into the United Kingdom had begun much earlier.

The first recorded introduction of a Giant Hogweed species in the UK comes from the seed list maintained by Kew Botanic Gardens, London, in 1817. This listed the receipt of *Heracleum giganteum*. In 1828, *Heracleum persicum* was noted in Cambridgeshire, but this time as a wild population. These plants were later re-considered by the foremost expert in the Hogweed group, Ida Mandenova, and re-classified as *Heracleum mantegazzianum*.

Within years of these initial records, the plant soon appeared in places as distant as Edinburgh and Bristol. Widespread distribution was brought about by enthusiastic collectors and gardeners who were keen to send and receive seeds. One shining example, John Henry Louden, comments:

"We do not know of a more suitable plant for the retired corner of a churchyard, or for a glade in a wood; and we have, accordingly, given one friend, who is making a tour in the north of England and Ireland, and another, who is gone to Norway, seeds for depositing in proper places." *The Gardener's Magazine*

The fashion for planting Giant Hogweed continued for most of the 19th century and, as seeds became more easily and cheaply available, the plant became more common in smaller private gardens. Horticultural interest declined after warnings about the health hazard and invasiveness of the plant were written about towards the middle and end of the 1900s. By this time the plant had been 'jumping the fence' for many decades and had escaped into river systems and watercourses. The rapid rate of spread of Giant Hogweed from its introduction is illustrated in Figure 3.2 and shows that since the 1960s the increase in the number of locations at which it has been recorded has been exponential.

Giant Hogweed growing in native range

Figure 3.1 Location of the Caucasus Mountains

Today, Giant Hogweed is widespread across the whole of the UK (Figures 3.2 and 3.3) but is rarely, if ever planted by gardeners. Instead, its main method of spread is along watercourses, railway, roads, and in the movement of soil contaminated by seeds or dumped garden debris. These are the habitats in which it is most commonly found, and with its garden origins and its conduits of spread closely linked to human populations, it is little surprise that it is also prevalent in urban areas.

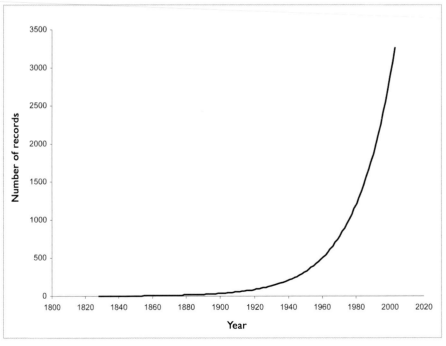

Figure 3.2   *Rate of spread of Giant Hogweed in the UK and Ireland*

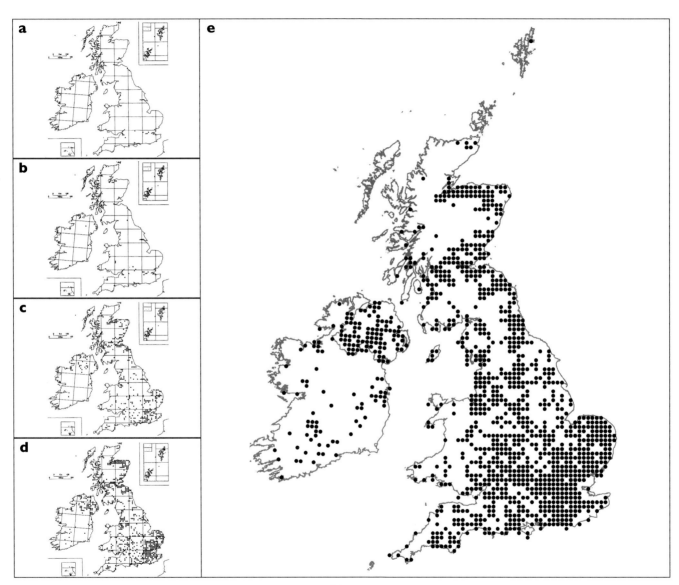

Figure 3.3 Giant Hogweed distribution in the UK and spread since 1920

Giant Hogweed records up to a,1920 b,1940 c,1970 d,1995 e, 2005. Source Biological Record Centre, Monks Wood and other data sources for e.

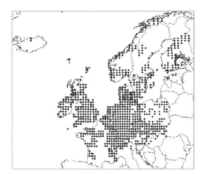

Heracleum mantegazzianum

Heracleum persicum

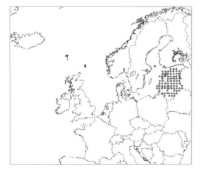

Heracleum sosnowskyi

Figure 3.5   *Distribution of Giant Hogweeds in Europe (from Nielsen et al. 2005) Chapter 12, number 19*

The other Giant Hogweed species

Heracleum sosnowskyi occurs naturally in the Caucasus and Transcaucasus as well as northeast Turkey (Figure 3.4). It was originally described in 1944 and was developed as a crop for northwest Russia from 1947 onwards due to its ability to thrive in a cold climate. Its large biomass was silaged to provide fodder for livestock. From the 1940s onwards, it was introduced to Latvia, Estonia, Lithuania, Belarus, Ukraine and the former German Democratic Republic, but is not present in the UK (Figure 3.5). Plantation schemes were eventually abandoned in the Baltic States, partly because the anise scented plants affected the flavour of meat and milk from the animals to which it was fed and partly because of the health risk to humans and cattle. However, agricultural production continued in parts of northern Russia.

Heracleum persicum originates from Turkey, Iran and Iraq (Figure 3.4). Its taxonomy in Europe is less clear than that of the other species, partly because it was the earliest to be described (1829) and some of the subsequent identifications of plants as *Heracleum persicum* were probably of *Heracleum mantegazzianum*. The only known wild populations of the plant *Heracleum persicum* in Europe are from Scandinavia (Figure 3.5), where the 'Tromsø Palm' and *Heracleum laciniatum* are alternative names for this species may occur rarely in the UK.

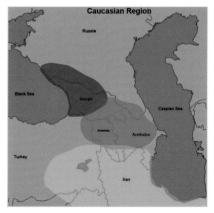

Figure 3.4   *Native range of Giant Hogweed Dark green – Heracleum mantegazzianum Light green – Heracleum sosnowskyi Yellow green – Heracleum persicum*

4. Biology, ecology and seed dispersal

Flowering and seed production

Understanding the biology and reproduction mechanisms of Giant Hogweed is useful for developing control methods and management strategies for the plant. It can live for several years but dies after the first time it bears seeds (Box 4.1). Leaves and stems persist in a rosette stage for the first few years and then, usually between the third and fifth year, it flowers. Under unfavourable conditions, such as on nutrient poor, shaded or dry sites or when regularly grazed, flowering is postponed until sufficient reserves in the taproot have been accumulated. Given such conditions, plants can live for at least 12 years.

The plant does not reproduce vegetatively (that is it cannot regrow from sections of stem or root) and relies exclusively on reproduction by seed. Flowers are arranged in compound umbels (Figure 4.1) and are composed of insect-pollinated flowers that have both male and female parts in the same flower (Chapter 12, number 25). Each umbel is composed of thousands of tiny white flowers (Figure 4.2). The pollen grains in a flower mature before the female reproductive structure becomes receptive to fertilisation. Therefore, seeds normally result from fertilisation between two plants (out crossing), but as there is an overlap in the male and female phases, this makes self-fertilisation possible. Seeds produced by self-pollination are viable; more than half of them germinate and give rise to healthy seedlings. This means that even a single isolated plant is capable of founding a new population, for example, as a result of a long distance dispersal event.

Box 4.1 Life history strategy

- Reproduces only by seed
- Lives for between 3-5 years before flowering
- Dies after setting seed
- If prevented from setting seed can live for up to 12 years
- Flowers appear between June and August
- Seeds usually set between July and August and released from late August
- Self-fertilisation is possible and so one plant can start a new invasion

Giant Hogweed seeds

Figure 4.1 Compound umbels of a Giant Hogweed plant

Figure 4.2 White flowers on Giant Hogweed umbel (Image from Šárka Jahodová)

Chapter 4 (including Boxes 4.1, 4.2 and 4.3) is adapted from Chapters 4 - 6 in Nielsen et al 2005, (Chapter 12)
Information contained here is based on studies from a range of European countries

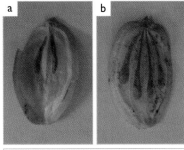

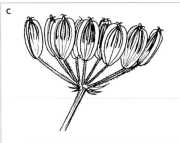

Figure 4.3 Giant Hogweed seeds showing oil ducts from different sides (a+b) and position on seed head (c) (Drawing from J. Schou)

In the UK, plants flower from June to August and seeds are released from late August to October. What appears to be the seed, is technically the fruit, and is actually made up of two paper-thin seeds stuck together back-to-back (Figure 4.3). For simplicity, in this manual the unit of reproduction is termed 'seed' (instead of the technical term mericarp). An average plant bears about 20,000 seeds (almost half of them on the terminal umbel), but individual plants with up to 50,000 seeds have been reported (Box 4.2). Although some of the seeds never germinate, the reproductive potential of the plant is enormous (Chapter 12, number 25).

Seed bank and germination

After falling from the parent plant, the seeds accumulate and mature in the soil (Chapter 12, number 14). The majority of seeds (95%) are concentrated in the top 5 cm of the soil (Box 4.2). In autumn, under dense stands, there can be as many as 12,000 living seeds/m² (on average 6,700) (Box 4.3). Some seeds are dead from the outset, some die and decay during the winter, yet there are on average still more than 2,000 living seeds/m² in the following spring, the vast majority of which are ready to germinate because they have lost dormancy over winter.

Giant Hogweed seeds will not germinate until their dormancy is broken. This is defined as a state in which the seed does not germinate although external conditions are suitable. The seeds remain buried in the soil until the dormancy is broken. These seeds are known as the seed bank and the process allows the seeds to wait for favourable conditions and hence reduces the mortality of the newly emerging seedlings. Cold and wet conditions are required in order to break dormancy and a period of two months at 2-4°C is sufficient to break under experimental conditions. In the field, dormancy is broken during autumn and winter (Chapter 12, number 16).

Box 4.2 Seed bank persistancy

- Average number of seeds per plant is 20,000, up to 50,000 have been recorded
- 95% of seeds are found in the upper 5cm of soil
- A significant cold period is required for germination
- In the first season 2,000/m² seeds are likely to germinate from one plant
- The majority of seeds only persist for 1-2 years (Box 4.3)

During spring germination, the short-term persistent seed bank is largely depleted due to germination and in summer it only contains about 200 living seeds/m². These remain dormant and about 8% are found to survive in the soil for more than one year; about 5% survive for two years after release from the parent plant (Box 4.3).

The maximum length of time seeds can survive in the soil seed bank is not known and has only been inferred from indirect evidence. The length of time can be reliably assessed only experimentally by burying seeds and following their fate over time. Nevertheless, the fact that at least a small fraction of seeds survive for at least two years is crucial for the course of invasion and application of adequate control measures. Given the high capacity to produce offspring, a single plant germinating from a single seed in the seed bank could start a new invasion.

After dormancy is broken, seeds germinate very easily (about 90% of the total seeds produced the previous year germinate under laboratory conditions at 8-10°C). In the field, seedlings reach high densities of up to several thousand/m² early in spring (March to April). The seedlings change the form of their leaves as they develop (Figure 4.4). The small size and round form of the initial leaves can make the plant difficult to recognise at this early stage (see Chapter 2).

Box 4.3. Density of living seeds found in the soil seed bank over time

Year	Average seed density (living seeds per m²)
Year 1 [1]	6,700
Year 2	200
Year 3	16
Year 4	10
Year 5	<10

[1] seeds fall from plant

Numbers assume that there are no contributions to the seed bank from other Giant Hogweed plants after Year 1

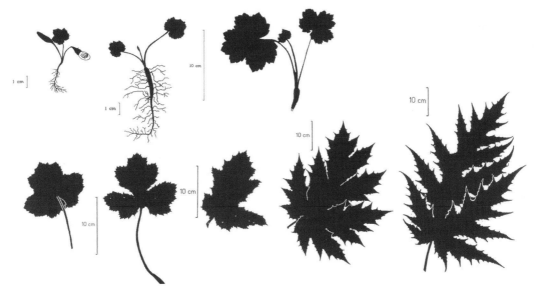

Figure 4.4 The growth phases of Giant Hogweed (from Ochsmann 1992, Chapter 12, number 20)

Although 98% of seedlings die in competition with each other during the process of self-thinning and due to shading by adult plants, the surviving plants in the following years create populations of plants with large leaf rosettes with almost complete ground cover. Rapidly developing Giant Hogweed populations shade out other plant species and Giant Hogweed often attains dominance in invaded sites. On average about 10% of plants flower and complete their life cycle, while the remainder survive in the rosette stage to flower the next year.

Seed dispersal

Giant Hogweed disperses only by seed and not from vegetative regrowth (for example sections of stem or root). Dispersal mechanisms include wind disturbance, dispersal along watercourses and accidental transportation by humans. A summary of dispersal mechanisms is provided in Box 4.4.

Giant Hogweed seeds are large and relatively heavy compared to those designed to be dispersed long distances by wind. As a result, the distance seeds are blown directly from the seed head is relatively small, the majority falling within 5m of the plant and the rest up to a distance of approximately 10m. However, this does not discount further spread by wind as the seeds are capable of being blown along the ground, especially where the ground is smooth and flat. This is particularly the case where passing traffic on roads and railways causes strong wind turbulence and on occasions when the ground is covered by snow or ice.

Giant Hogweed is most commonly spread along watercourses. Seeds have been known to travel for 3 days in the water channel, travelling over 1km. They easily enter the watercourse having been blown from plants growing along the riverbank or, during late autumn and winter, by plants that have been blown over and fallen into the watercourse. Seeds come to rest in eddies or are caught in vegetation along the river and easily take hold and germinate. More rarely, flooding events can occur that spread seeds further than the river channel. These events can spread seed across the floodplain, and in some cases can be responsible for allowing seeds to travel upstream, as well as downstream.

Human facilitated spread is also common for Giant Hogweed. Historically this has included deliberate planting by gardeners (see Chapter 3); however spread is primarily accidental, along corridors, for example railways and roads (Figure 4.5), or through transport of contaminated material, for example flytipping or reuse of soils for landscaping.

Figure 4.5 Giant Hogweed growing in central reservation of a motorway

Box 4.4 Seed dispersal mechanisms

Dispersal mechanisms	Frequency	Approx. distance travelled	Habitats commonly affected
Wind – falling from plants	Very frequent	0 – 10m	Any habitat within 10m of plants
Wind - skimming along ground	Moderately frequent	At least 50m	Habitats connected by flatsurfaces (e.g. roads, flat dry ground, ice / snow in winter)
Water courses – spread downstream	Very frequent	Over 1km	Habitats downstream, particularly where river riverbanks have been disturbed or left unmanaged
Water courses – flooding	Moderately frequent	Extent of flooding	Flood plain
Railway-lines	Frequent	0-100m	Railway line embankments and habitats associated with them
Roadways	Frequent	0-100m	Roadways and habitats associated with them
Mowing of grassy areas	Frequent	At least 50m	Grassy areas, parks, recreation grounds
Deliberate planting	Rare	UK wide	Amenity areas and private land including gardens, farm, parkland, churchyards and country estates
Dumping of soils	Moderately frequent	UK wide	Areas prone to flytipping including brownfield sites. Landfill sites
Transfer of contaminated soils between sites	Moderately frequent	UK wide	Development sites, landscaped areas

Biological and ecological characteristics of invasiveness

The main biological and ecological characteristics of Giant Hogweed that make it such a successful invader are summarised below. These features, together with efficient seed dispersal by human activities, water and wind, give Giant Hogweed enormous invasion potential.

- Germination in early spring before the resident vegetation appears;

- Low mortality of plants once they become established;

- Fast growth of rosettes allowing rapid development of populations and the ability to form dense cover and place leaves above the resident vegetation;

- A stable proportion of plants that flower and produce seeds;

- Ability of plants under stressful conditions to postpone flowering until a time when sufficient reserves are stored;

- Flowering sufficiently early in the year in order to ensure that the development of seeds is complete;

- Ability of self-pollination leading to production of viable seed;

- High fecundity allowing a single plant to start an invasion;

- High density of seeds in the seed bank, with some seeds surviving for at least two years and some for longer (Box 4.3);

- Efficient breaking of dormancy by cold temperature during autumn and winter; and

- Very high percentage of germination regardless of which umbel of the parent plant the seeds are produced from.

Giant Hogweed growing next to North Circular, London

5. Impacts

Habitat and biodiversity loss

Giant Hogweed is one of the most competitive plants in the flora of the UK. It starts to grow early in the year which, coupled with its height, allows it to outgrow other species. The plant's large leaves then shade out other potential competitors. It produces huge number of seeds, a large proportion of which persist in the seed bank (Chapter 4), ensuring that there are always plenty of competitors in the spring when germination occurs. In addition, the rapid dispersal of Giant Hogweed seeds by humans, wind and water ensure that it is widely distributed. The result is that Giant Hogweed tends to dominate wherever it grows, pushing out native species.

While the impact on biodiversity may be relatively small where Giant Hogweed grows in urban areas and on brownfield sites, the effect is much more significant where it grows on riverbanks. These are often some of the most ecologically sensitive and vulnerable habitats in the UK and the majority of large-scale Giant Hogweed control operations in the UK have been introduced to help stop, and hopefully recover, damage to important habitats such as these (Box 5.1).

Box 5.1 Examples of impacts of Giant Hogweed on sites in the UK

River Tweed Catchment Area, Scottish Borders

- Threatening the value and biodiversity of a designated Site of Special Scientific Interest (SSSI) and a Special Area of Conservation (SAC)
- Posing a threat to the high profile salmon fishing for which the river is famous
- Preventing amenity use of the river by canoeists, boat users and walkers
- Causing severe riverbank erosion
- Posing a health hazard to the general public

Tamar Valley, Cornwall / Devon border

- Degrading the aesthetic value of an Area of Outstanding Natural Beauty (AONB)
- Out-competing and suppressing the growth of native flora

River Usk, South Wales

- Infesting approximately 40km of riverbank
- Causing access problem for anglers along river renowned for salmon and trout populations

---

Figure 5.1 Giant Hogweed control leaving a river bank exposed

River bank erosion and flood risk

Giant Hogweed commonly forms dense stands on riverbanks. When natural die-back occurs in winter or control is undertaken, the riverbank is left devoid of any vegetation and exposed to erosion over the late autumn and winter period (Figure 5.1). In addition, large holes are left where the plant taproots had been growing. These can destabilise the riverbank, increasing the possibility of erosion, and in some cases allowing whole sections to be washed away. This effect is most noticeable after herbicide control, where all of the taproots have been killed as opposed to just those plants that have flowered and died in a season.

When water levels in a river rise during flood conditions, the increased mass of plant material along the riverbank caused by Giant Hogweed can impede the river flow and increase the likelihood of flooding.   Additionally, dense stands obscure the river making access to the bankside difficult or impossible and preventing river managers from inspecting it.

Health risk

Giant Hogweed poses a serious threat to human health. The clear, watery sap, found in all parts of the plant contains varying amounts of photosensitising compounds called furanocoumarins (also called furocoumarins). These compounds can cause painful burning and blistering on contact with human skin. This effect is further enhanced by exposure to sunlight.

After contact, a red rash and blistering, occasionally severe, can occur within 24-48 hours (Figure 5.2). The skin starts to discolour, turning a red or dark purple colour, which can last for several months or even years. Victims are unaware of the damage that is being done, as touching the plant is painless, and hence they may continue to remain in contact with the plant. For the same reason, victims often do not recognise or report the cause of symptoms (Box 5.2).

Box 5.2 A more common problem than you might think

Incidents of injury by Giant Hogweed are not uncommon, but are often under reported. One account reported in a local newspaper was of a six year-old boy who had fallen into a clump of plants while playing on council land in Scotland. He had to be hospitalised immediately and was treated for severe burns on his stomach and back. A spokesman from the hospital commented that they expected to deal with two cases of Giant Hogweed burns every week during the summer.

Reactions vary between individuals with some being hypersensitive to the plant's sap and others only suffering a mild response. Once skin has been affected, it may remain sensitive to ultraviolet light for many years and sun block may be required in even normal light conditions. Contact with the eyes can cause temporary blindness and, in a minority of cases, permanent blindness. Box 5.3 provides recommendations for action should a person come in contact with the sap of Giant Hogweed.

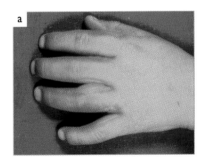

Box 5.3 Action to take if contact with Giant Hogweed occurs

- Keep any areas of skin that have come into contact with sap in dark conditions, away from sunlight
- As soon as facilities can be reached, wash the skin in soapy water
- If Giant Hogweed sap comes into contact with the eyes, immediately wash them with eyewash solution
- Seek medical advice after contact with Giant Hogweed sap, especially after contact with the eyes

---

Those most at risk from Giant Hogweed are children who play in areas where Giant Hogweed can be found (Box 5.4), and find the hollow stems an attractive toy employed as pretend telescopes, peashooters or swords. The result is often painful blistering on the body and skin around the eye. While only relatively minor injuries have been reported to date in the UK, more severe injuries and even fatalities have been reported from Eastern Europe.

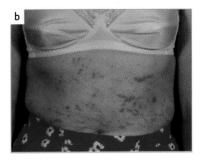

As the plant is often spread by humans and is capable of growing in the urban environment, it is common to find Giant Hogweed growing in conflict with people, such as in parks and recreation grounds, along riverside paths, on footpaths and roads as well as private gardens and council owned property (Figure 5.3). Box 5.5 provides guidance for action to take if Giant Hogweed is growing on your property.

Box 5.4 Children are particularly at risk from injury

"Anger and pain as plant inflicts burns on girls" *The Kent Messenger 2006*

Children are often the most at risk from Giant Hogweed as it is commonly found in parks and other areas where they are likely to play. In this instance, reported in a local newspaper, two young girls were seriously burned after coming in contact with the plant while walking in the countryside.

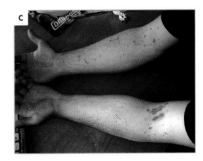

Figure 5.2 Skin damage caused by Giant Hogweed (image from Mark Pritchard).

Figure 5.3 Giant Hogweed growing through a park bench east London

---

Box 5.5 Action to take if Giant Hogweed is found on your property

---

- Make sure all those that could come into contact with the plant are warned of the dangers. If on or near public land, signage should be erected to warn of the dangers
- Keep pets away from the plant as sap can be carried to humans on animal fur
- Erect a fence or other barrier to restrict access to the plant
- Once all of these have been done, a control strategy should be implimented to eradicate the plant

---

Impacts on amenity and local economy

Giant Hogweed creates a variety of problems for land managers in amenity areas. Dense stands of the plant can block access to riverbanks and lakes, and create difficulties for those that need access to them, for example river managers and anglers. Where both banks of a river have been infested with dense stands of Giant Hogweed, safe access to the water is impossible. Footpaths along riverbanks may also be rendered unusable for walkers and runners by the presence of the plant.

Instability of riverbanks dominated by Giant Hogweed can also pose a threat to fish spawning habitats and fish stocks. Dense stands of Giant Hogweed have developed along some of Ireland's popular salmonid and coarse fishery rivers, for example the River Dee in County Louth (Chapter 12, number 40). The subsequent winter erosion has resulted in large quantities of soil being washed into the river. Fine soil and silt particles can clog the spaces in-between riverbed sediments where salmonids prefer to spawn and make the channel more favourable for abundant aquatic plant growth. Salmon and trout eggs laid in gravel can only properly develop when a current of water is passing through the gravel and can be smothered by fine sediment. Through this, and the altering of the characteristics of the riverbed, Giant Hogweed has the ability to damage productive salmonid fisheries.

Health and safety, cost and delay to development

Giant Hogweed frequently occurs on brownfield development sites (Figure 5.4). In these circumstances, there is a duty of care to ensure safety precautions are in place to prevent injury during the development of the site. The client, lead contractor and site manager should adopt appropriate safety procedures as soon as they are aware that Giant Hogweed is present on their land (Box 5.6).

Box 5.6 Measures to take if Giant Hogweed is found on a development site

- Immediately erect semi-permanent fencing around the plant (Figure 5.4)
- Install signage warning personnel of the dangers and advising unauthorised personnel to keep clear
- Brief relevant personnel of the presence of Giant Hogweed during site inductions
- Ensure the first aid officer is aware of the risk and the methods of treatment
- Place an information notice in site office, including instructions for treatment if accidental contact occurs
- Ensure washing facilities, including eyewash, are available near to where contact with the plant may occur
- Use a control strategy to eradicate the plant (Chapter 10)

Figure 5.4 Fencing off Giant Hogweed on a brownfield site

Where Giant Hogweed grows on a development site it should be eradicated, not only to remove the risk to human health, but also to prevent an offence under legislation (see Chapter 6). The cost of eradicating the plant must be met by the developer and can be high if there is only a short time frame for control (Box 5.7). The risk of delay and cost to development can be reduced if Giant Hogweed is identified early on. Many developers now require land to be surveyed for invasive species such as Giant Hogweed and Japanese Knotweed even before the land is purchased.

Box 5.7 A serious problem for development

"Hogweed may halt Olympic Development" *This is local London 2005*

The east London area of the 2012 Olympic Games suffers from a significant invasion of Giant Hogweed, as well as other non-native plant species. Many London newspapers reported throughout the development for the Games the likely problems that would be encountered in dealing with these plants. With a restricted timescale for development and over 18 hectares of invasive plants including Japanese Knotweed and Giant Hogweed to eradicate, the estimated cost for control of all the invasives is over £100m.

Giant Hogweed growing on a footpath in east London

6. Legislation

In the United Kingdom, legislation concerning Giant Hogweed is provided by the following instruments (summarised in Box 6.1):

Wildlife and Countryside Act 1981 (Schedule 9, Section 14(2))

This legislation covers England, Scotland and Wales, with the Wildlife (Northern Ireland) Order 1984 providing equivalent legislation for Northern Ireland. The Act and the Wildlife Order make it an offence to "plant or otherwise cause to grow in the wild" any species listed on part II of Schedule 9 of the act. Giant Hogweed is one of the two terrestrial plants species listed (the other is Japanese Knotweed). In Scotland, the Nature Conservation Act 2004 amended Schedule 9 to include hybrids of Giant Hogweed, however this does not apply to the rest of the UK.

The penalty for conviction under the Wildlife and Countryside Act is an unlimited fine and imprisonment for up to two years.

Natural Environment and Rural Communities Act 2006 and Nature Conservation (Scotland) Act 2004

Both of these instruments add to the Wildlife and Countryside Act 1981 giving the Secretary of State or Scottish Ministers the power to make it an offence to sell or advertise the sale of specified non-native species. In 2007 Giant Hogweed had not yet been listed but it is likely to be sometime in the near future. The Natural Environment and Rural Communities 2006 act applies to England and Wales only, while the Nature Conservation (Scotland) Act applies only to Scotland. Northern Ireland is not covered by either act.

This legislation also gives the relevant authority the power to provide or approve codes of practice relating to non-native species. Although these are not legally binding, they are admissible as evidence in proceedings and should be taken into account by judges where relevant.

Environmental Protection Act 1990 Section 33 / Section 34

Amongst other things, this act concerns the management and disposal of Giant Hogweed, which if taken away from its point of origin is considered to be a controlled waste. Controlled waste must be disposed of in an appropriate manner to prevent pollution or threat to human health. Giant Hogweed material must therefore be safely disposed of at an appropriately authorised landfill site and accompanied by appropriate documentation under the Duty of Care, described in section 34 of the Act.

The penalty for an offence under this legislation is imprisonment for up to two years and a fine, or both.

Other legislation

Other legal instruments may apply to Giant Hogweed where it is considered to be a pest, nuisance or public health hazard. These are not specifically targeted towards Giant Hogweed, but are general instruments. They include:

- Provision within common law to take civil action against neighbouring landowners where the spread of Giant Hogweed is considered to be a private or public nuisance, and

- Action taken by a local authority if Giant Hogweed is considered to be a statutory nuisance as set out in Part III of the of the Environmental Protection Act 1990.

Box 6.1 Summary of relevant legislation

- It is illegal to plant or otherwise cause Giant Hogweed to grow in the wild in the UK

- If taken away from the site of origin, Giant Hogweed waste must be disposed of at a landfill site that is authorised to accept it

- Giant Hogweed waste that is disposed of at a landfill site must be accompanied by appropriate waste transfer documentation

- Other instruments of general common law may apply to Giant Hogweed

*Giant Hogweed on the River Lee,
Hertfordshire*

7. Policy

There is global recognition that invasive species, including Giant Hogweed, pose a serious threat to biodiversity. The UN Conference on Alien Species (Trondheim 1996) recognised that invasive alien species were the second greatest threat to biodiversity, after habitat destruction. Global concern has resulted in international agreements to take action to help prevent the spread and minimise impacts of invasive species. These have been translated by UK governing bodies into policy at the national and regional level. Key international, national and regional policy and guidance governing invasive species, including Giant Hogweed, is outlined here. The agencies involved in their delivery in the UK are detailed in Box 7.1.

International guidance and policy relating to Giant Hogweed

Bern Convention 1982

The United Kingdom is a signatory of the Bern Convention, which aims to help conserve and protect all wild plants and animal species. As part of this, signatories agree to 'strictly control the introduction of non-native species'.

Convention on Biological Diversity 1992

The guiding international agreement for invasive species is the Rio Summit 1992, Convention on Biological Diversity (CBD). All participating nations, including the UK, are called on to 'prevent the introduction of, control, or eradicate those alien species, which threaten ecosystems habitats or species.' Guiding Principles for action on invasive alien species were developed which include: prevention of introduction, early detection, rapid control, containment and long-term control.

Ramsar Convention 1976

The spread of Giant Hogweed in the UK has become particularly apparent on riverbanks and waterways, as water provides a means for the plant to spread. The Convention on Wetlands (Ramsar Convention) calls upon Contracting Parties, including the UK, to address the environmental, economic and social impacts of invasive species on wetlands within their jurisdiction wherever possible. The Ramsar Convention urges Contracting Parties to prepare an inventory of alien species in wetlands within their jurisdictions, to establish programmes to target priority invasive species with a view to control or eradication and to provide database records for invasive species.

Box 7.1 The role of agencies with environmental responsibilities for Giant Hogweed

Department	Description
The Department for Environment, Food and Rural Affairs (Defra) Scottish Executive Environment and Rural Affairs Department (SEERAD) Department of the Environment (DOE)	Defra is the English and Welsh government department responsible for wildlife and therefore invasive non-native species including Giant Hogweed. In Scotland, SEERAD is responsible for these; in Northern Ireland, DOE has responsibility.
Environment Agency (EA) Scottish Environmental Protection Agency (SEPA) Environment and Heritage Service Northern Ireland (EHS)	Known as Environmental Protection Agencies (EPAs) the EA and SEPA are responsible for the management of arterial water courses such as rivers and streams which are often impacted by Giant Hogweed. and in delivering the EU Water Framework Directive and its implications for non-native species in UK water bodies. In addition, these agencies are responsible for the management of controlled waste, a classification relevant to Giant Hogweed in some situations. In Northern Ireland, waste regulations are enforced by the EHS.
British Waterways (BW) Waterways Ireland	In 2005, BW produced a Code of Practice that applies to all works undertaken on waterways that fall under BW jurisdiction. Section 4.33 of the Code refers to Giant Hogweed and other species that are 'a cause for concern' and need to be disposed of responsibly. The Code also states that a requirement for contractors to avoid the spread of Giant Hogweed must be inserted into all work contracts. BW have also overseen local control of Giant Hogweed where stands are present along waterway corridors. Waterways Ireland manages the navigable waterways in Ireland and has an environmental policy and code of practice which addresses the issue of non-native invasive species.
Natural England Countryside Council for Wales Scottish Natural Heritage Northern Ireland's Environment and Heritage Service	These are Statutory Nature Conservation Organisations that undertake management of Giant Hogweed where it has been recorded on statutorily protected nature conservation sites (e.g. SSSIs and SACs). The management of Giant Hogweed on these sites is important as large infestations of Giant Hogweed could affect the nature conservation value of the site and leave it in an unfavourable condition. The SNCOs have also supported initiatives and entered into partnerships with other conservation bodies and local authorities to ensure Giant Hogweed is controlled effectively over larger areas.
Internal Drainage Boards (IDBs)	IDBs are responsible for maintaining drainage and drainage ditches at a local level around the UK. Various IDBs around the UK have engaged in local partnerships with other drainage authorities (e.g. Environment Agency) to address the issue of Giant Hogweed.

National guidance and policy relating to Giant Hogweed

UK Biodiversity Action Plan

The UK Biodiversity Action Plan (UKBAP) is the UK government's response to the Convention on Biological Diversity and provides a framework for the protection and enhancement of biodiversity in the UK. Within the UKBAP invasive species are identified as a significant threat to a large proportion of key habitats and species.

Defra Review of Non-native Species Policy 2003 (Chapter 12, Useful Links)

Defra undertook a Review of Non-Native Species Policy in 2003 and produced several key recommendations to improve measures to limit the ecological and economic impact of these species in the UK. The Defra Review recognised that the arrangements in the UK for handling issues relating to non-native species were insufficient and recommended eight key areas for development, which are being progressed. The eight key recommendations to government were to:

1. Designate or create a single lead co-ordinating organisation to undertake the role of co-ordinating and ensuring consistency of application of non-native species policies across Government.

2. Develop comprehensive, accepted risk assessment procedures to assess the risks posed by non-native species and identifying and prioritising prevention action.

3. Develop codes of conduct to help prevent introductions for all relevant sectors in a participative fashion involving all relevant stakeholders.

4. Develop a targeted education and awareness strategy involving all relevant sectors.

5. Revise and update existing legislation to improve handling of invasive non-native species issues.

6. Establish adequate monitoring and surveillance arrangements for non-native species in Britain.

7. Policies should be established with respect to management and control of invasive non-native species currently present or newly-arrived in the wild, and operational capacity be developed to implement these policies.

8. Stakeholders should be fully consulted and engaged in development of invasive non-native species policies and actions through a mechanism such as a consultative forum.

Horticultural Code of Practice 2005 (Chapter 12, Useful Links)

Developed by Defra, Scottish Executive, Welsh National Assembly and stakeholders in the horticulture industry, this voluntary code is designed to provide guidance on the planting of non-native species by the horticulture industry.

Local / Regional policy and guidance relating to Giant Hogweed

Local Biodiversity Action Plans (LBAPs)

LBAPs are provided to help carry out the objectives of the UKBAP at a local level typically a county. One of the main aims of LBAPs is to help raise awareness of the natural environment among members of the local community and take action to improve it. As a result LBAPs can include raising awareness of Giant Hogweed and removing it from environmentally sensitive areas.

The planning process

There are various Local Authorities across the country, which are proactive in the eradication of Giant Hogweed and have attempted to control the species through the planning process. For example, if Giant Hogweed is recognised on a development site, a planning condition is applied which requires the applicant to submit a method statement, which details a program for the eradication of the Giant Hogweed from the site, prior to works commencing (see Box 7.2).

Protection of sites for local nature conservation interest

Across the UK there are areas afforded special status for their local level of conservation interest, for example County Wildlife Sites. In some cases groups such as the Wildlife Trust undertake to manage those sites. Many sites of conservation importance, especially those containing important watercourses, are threatened by the introduction and spread of Giant Hogweed. The action plans developed for these sites often include a programme for the control of invasive species, including Giant Hogweed.

Cross-sector partnerships

Across the UK several partnerships have been established to guide the development of strategies aimed at controlling Giant Hogweed and other invasive species. These often comprise representatives of a range of interested stakeholders including Local Authorities, private industry and statutory agencies such as the Environment Agency. It is recognised that effective large-scale control of Giant Hogweed, and other invasive species, can most effectively be achieved through coordinated action involving public and private bodies and landowners.

Box 7.2 Example of a planning condition to deal with Giant Hogweed

Full details of a scheme for the eradication of Giant Hogweed to be carried out by the developer, shall be submitted to and approved in writing by the Local Planning Authority, prior to the commencement of any development. This plan shall include a timetable for implementation. Should a delay of more than one year occur between the date of approval of the eradication scheme and either the date of implementation of the eradication scheme or the date of development commencing, a further site survey must be undertaken and submitted to the Local Planning Authority for approval in order to ensure that the agreed scheme is still applicable.

Giant Hogweed on the River Brent, London

8. Developing a management strategy for controlling or preventing an invasion

Stages of a project

Before embarking on a Giant Hogweed management project, planning is essential, including determining clear aims and objectives, raising support and monitoring the success or otherwise of the project. This section describes the main stages to consider (Figure 8.1). It may not be necessary to undertake them all, and a project could be just one stage or a number of stages. The advice given is applicable both to projects dealing with infestations and those that aim to prevent an invasion occurring.

Setting the scope of the project

Establishing aims and objectives

The aim of the project should be clearly identified, straightforward and deliverable. In smaller areas this could be the eradication of Giant Hogweed, on a larger scale, control and long-term management are often more realistic goals. It could also be to prevent Giant Hogweed invading into an area. Aims should be closely linked to, and attainable within the budget of the project. Projects and strategies cut-short through lack of funding frequently have little impact because controlled areas re-colonise and the effort is wasted. However, even with a relatively small budget a well-planned strategy can prove useful, for example tackling a Giant Hogweed stand at the top of a watercourse that would otherwise spread. The objectives should be measurable with agreed milestones set. Specific objectives should also be set to measure progress and achievement throughout the project. Such objectives might be to produce and gain approval for a policy, raise awareness of colleagues in your organisation, and run a training programme for staff applying herbicides.

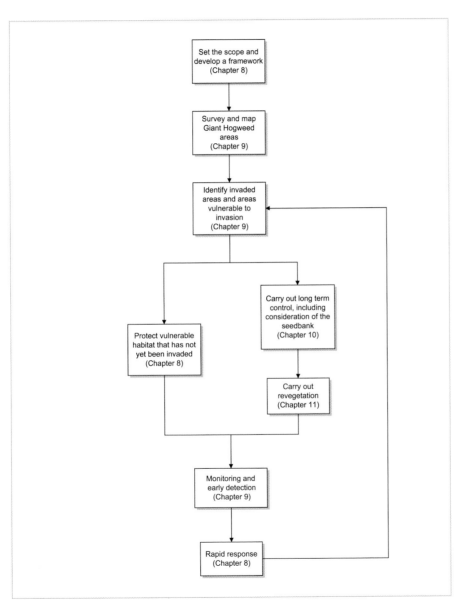

Figure 8.1   Stages of a management strategy

Establish an appropriate timescale

The timescale of a project should be chosen to closely match the aims and objectives set out for it. While small-scale projects may have a defined start and end point, larger scale projects often rely on continual management in order to achieve their aims. Preventing the invasion of an area, for example, requires on-going management. Unfortunately, funding is often difficult to secure for long-term projects and so re-application for budget extensions may have to be carried out to keep the overall project going. In the long term, costs will decrease as the impact of the project takes effect. Efforts should be made to find an appropriate budget or organisation to take over once the main development has been completed. This should include control measures or plans to monitor for invasion or re-invasion, thereby ensuring sustainable management.

Identify the project area

The project area should be carefully selected, that is the area within which control is proposed or the area from which it is planned to exclude Giant Hogweed. It could range from a garden or park to river catchment, county, region or even a country. The area is usually defined by two factors: human administrative boundaries (for example private land, town, county, country) and, or natural divisions (for example nature reserve, valley, river catchment). Each has benefits and drawbacks (Box 8.1).

Box 8.1 Advantages and disadvantages of project areas defined by human and natural boundaries

Definition of area	Advantages	Disadvantages
Human administrative areas (for example towns, counties, regions)	• Well known and understood areas • Groups, societies and organisations are used to working with these boundaries • Administration already in place. Clear leaders / responsibility can be identified	• Often artificial boundaries without regard for natural divisions • May form only part or parts of a larger ecological system which then can be a source of re-invasion
Natural divisions (for example water catchments areas)	• Encompasses an ecological system and therefore should be more resistant to natural re-invasion • Management can be undertaken on an ecologically logical basis, for example from top of catchment downstream	• Can encompass many different authorities, landowners, and agencies who must be made to work together • Often not well known or easily defined boundaries • Responsibility often shared between several administrations • Lack of clear leadership

A proven approach for large-scale management is to provide control across a water catchment. This has the benefit of being a relatively contained natural system, and for Giant Hogweed, a river corridor is the most likely source of natural spread.

Administrative areas have the benefit of well-recognised delineation, and existing structures and organisations to co-ordinate with, for example local authorities and wildlife groups. However, these areas are potentially vulnerable to invasion from natural sources outside their boundaries (for example rivers entering but not originating in the project area).

Small-scale sites should be considered in the wider context in relation to the potential for invasion or reinvasion. In this situation consideration should be given to controlling corridors of reinvasion back into the site, for example rivers, railways, roads and flytipping, as well as plants in the immediate site environs.

While the choice of management area will be dictated by the aims and scope of the project, thought needs to be given to co-ordinating the relevant people and organisations, and how to manage the potential for natural sources of re-invasion.

Areas that do not have Giant Hogweed but are vulnerable to invasion can be identified by undertaking a risk assessment. This allows resources to be efficiently targeted where they will be of most use and this can be done by surveying pathways through which invasion can occur. Sites within dispersal distance of Giant Hogweed plants are particularly vulnerable and areas with suitable habitat are also at risk (for example a good water supply, high sunlight, recently disturbed and no active land use).

Developing a framework

Engaging stakeholders

The success of a project will depend in large measure on the support and cooperation of a range of people and organisations. Key stakeholders should be identified early on and could include colleagues in your organisation, local authorities, landowners, businesses and industry and the statutory agencies such as the Environment Agency, Scottish Environmental Protection Agency or Environment and Heritage Service (Figure 8.2).

Figure 8.2 Key stakeholders engaged in Giant Hogweed control on the River Medway. (Image from Medway River Project, Chapter 12, Useful Links)

Consultation

Initial consultation is a critical stage of engaging stakeholders and ensuring that key groups feel empowered to express their views. Stakeholders should consult in a structured way to bring together a framework for the project. Key consultation topics could include identifying existing expertise and resources, exploring potential for sharing responsibilities, establishing policy and raising funding. Consultation can be undertaken on a one-to-one basis, through e-mail and meetings, and, or by establishing a working group. Support could be gained from the local community through leaflets and other material to inform them of the project and how to contribute (Figure 8.3).

During the operation of the project, consultation is a key tool to ensure appropriate progress is being made. Key stakeholders should be engaged in repeat meetings, and if applicable open forums could be held to discuss the progress that is being made and canvas further ideas and support. Open consultation may help bring in additional stakeholders and allow all individuals to express their views.

Setting policy

A sound and sustainable way of delivering a project is to develop and agree a policy that incorporates the key elements such as aims and objectives, responsibilities, funding and measures for success (Box 8.2). Such a policy to manage Giant Hogweed could be embedded in a local authority Local Development Framework, developed and accepted by a joint working group or forum, or simply a policy internal to your organisation. Such a policy needs to be used to guide and review project progress and the policy itself should be revised on a regular basis.

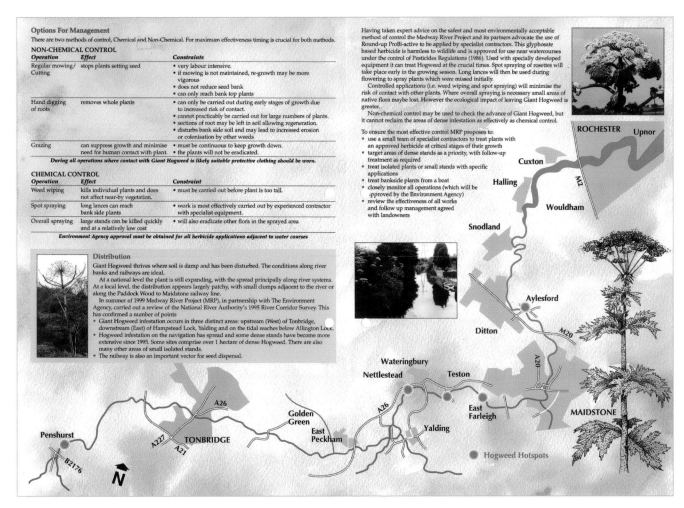

Options For Management

There are two methods of control, Chemical and Non-Chemical. For maximum effectiveness timing is crucial for both methods.

NON-CHEMICAL CONTROL

Operation	Effect	Constraints
Regular mowing/ Cutting	stops plants setting seed	• very labour intensive. • if mowing is not maintained, re-growth may be more vigorous • does not reduce seed bank • can only reach bank top plants
Hand digging of roots	removes whole plants	• can only be carried out during early stages of growth due to increased risk of contact. • cannot practicably be carried out for large numbers of plants. • sections of root may be left in soil allowing regeneration. • disturbs bank side soil and may lead to increased erosion or colonisation by other weeds
Grazing	can suppress growth and minimise need for human contact with plant.	• must be continuous to keep growth down. • the plants will not be eradicated.

During all operations where contact with Giant Hogweed is likely suitable protective clothing should be worn.

CHEMICAL CONTROL

Operation	Effect	Constraint
Weed wiping	kills individual plants and does not affect near-by vegetation.	• must be carried out before plant is too tall.
Spot spraying	long lances can reach bank side plants	• work is most effectively carried out by experienced contractor with specialist equipment.
Overall spraying	large stands can be killed quickly and at a relatively low cost	• will also eradicate other flora in the sprayed area

Environment Agency approval must be obtained for all herbicide applications adjacent to water courses

Distribution

Giant Hogweed thrives where soil is damp and has been disturbed. The conditions along river banks and railways are ideal.

At a national level the plant is still expanding, with the spread principally along river systems. At a local level, the distribution appears largely patchy, with small clumps adjacent to the river or along the Paddock Wood to Maidstone railway line.

In summer of 1999 Medway River Project (MRP), in partnership with The Environment Agency, carried out a review of the National River Authority's 1995 River Corridor Survey. This has confirmed a number of points

• Giant Hogweed infestation occurs in three distinct areas: upstream (West) of Tonbridge, downstream (East) of Hampstead Lock, Yalding and on the tidal reaches below Allington Lock.
• Hogweed infestation on the navigation has spread and some dense stands have become more extensive since 1995. Some sites comprise over 1 hectare of dense Hogweed. There are also many other areas of small isolated stands.
• The railway is also an important vector for seed dispersal.

Having taken expert advice on the safest and most environmentally acceptable method of control the Medway River Project and its partners advocate the use of Round-up ProBi-active to be applied by specialist contractors. This glyphosate based herbicide is harmless to wildlife and is approved for use near watercourses under the control of Pesticides Regulations (1986). Used with specially developed equipment it can treat Hogweed at the crucial times. Spot spraying of rosettes will take place early in the growing season. Long lances will then be used during flowering to spray plants which were missed initially.

Controlled applications (i.e. weed wiping and spot spraying) will minimise the risk of contact with other plants. Where overall spraying is necessary small areas of native flora maybe lost. However the ecological impact of leaving Giant Hogweed is greater.

Non-chemical control may be used to check the advance of Giant Hogweed, but it cannot reclaim the areas of dense infestation as effectively as chemical control.

To ensure the most effective control MRP proposes to:
• use a small team of specialist contractors to treat plants with an approved herbicide at critical stages of their growth
• target areas of dense stands as a priority, with follow-up treatment as required
• treat isolated plants or small stands with specific applications
• treat bankside plants from a boat
• closely monitor all operations (which will be approved by the Environment Agency)
• review the effectiveness of all works and follow up management agreed with landowners

Figure 8.3 Example of a Giant Hogweed information leaflet used to engage local support in the Medway Valley (Image from Medway River Project Chapter 12)

Coordination

A combination of engaging stakeholders, consultation and setting policy assisted by well thought out and sustained communication and clear responsibilities will help to ensure good coordination. Important considerations within this are the use of existing structures where possible and an agreement to share resources including data. A vehicle that has proved an efficient and successful means of achieving coordination has been the establishment of a forum to run the project. Examples of forums established to deal with non-native invasive species area The Cornwall Knotweed Forum, the Tweed Forum and the Swansea Japanese Knotweed Forum (see Chapter 12, Useful Links).

Box 8.2 Examples of how different organisations have used local / regional policy instruments and stakeholder engagement to manage Giant Hogweed in the UK

Stirling Council has implemented a strategy for the eradication of Giant Hogweed. Working with Stirling Landfill Trust and Scottish Natural Heritage, Stirling Council is attempting to eradicate Giant Hogweed where it is present in areas within the Council's jurisdiction. Stirling Council provides advice on Giant Hogweed to the public and offer assistance to householders who have Giant Hogweed present in their gardens. Spraying is also undertaken on Giant Hogweed when members of the public request it.

Medway Valley Countryside Partnership is sponsored by the Environment Agency and several Local Authorities and has successfully fostered the co-operation of more than 70 landowners and has substantially reduced the extent of Giant Hogweed throughout the River Medway catchment area.

West Country Rivers Trust has helped to raise awareness of the impacts of Giant Hogweed by offering training in the identification and control of Giant Hogweed and other invasive species. WCRT have also hosted a Symposium on invasive plants in partnership with Cornwall Knotweed Forum.

Cornwall Knotweed Forum was established to help prevent the spread of Japanese Knotweed and to develop good practice techniques to assist in the control of the plant. Although targeted at Japanese Knotweed, the approach that the Forum has taken could also apply to Giant Hogweed control. The Forum recognises that no single organisation is responsible for management of Japanese Knotweed and therefore responsibility for dealing with the plant in the county resides with a range of organisations. Various organisations support a Forum member with a good knowledge of Japanese Knotweed issues. Education is the most important role of the Cornwall Forum and its activities have resulted in a considerable improvement in public awareness within the county. Similar strategic partnerships to address the issue of invasive species have also been established in Swansea, Merseyside and Devon.

Identifying a champion

The success of a project is often dependent on a champion having the commitment and determination to deliver the project aims. A champion can be either an organisation willing and able to drive the project, typically also contributing funding or obtaining funding, or an individual with the personality, skills and energy to coordinate and motivate those involved in the project. Whilst an organisation might be persuaded to take on the role of champion, the mantle is usually taken up of necessity, for example due to a statutory responsibility, sometimes coupled with a member of staff keen to combat the problem. Where the champion is an individual, he is normally appointed using funding obtained specifically to deliver the project.

Finance and funding

A key factor, if not the key factor, is achieving funding for your project. Sources can range from existing resources from one organisation or between organisations or through regional, national or European Union support. The funding might only cover a specific facet of your project, for example eradication of Giant Hogweed in only a part of your project area, and it may be useful to approach more than one source. Securing initial funding can help secure further support and in-kind funding can be very valuable in successfully applying for matching funding. Examples of funding sources available in the UK include the landfill tax credits scheme and the heritage lottery fund.

Figure 8.4 Raising awareness using the media

Raising awareness

Awareness raising and the provision of information is an important part of a management strategy to deal with Giant Hogweed (Box 8.3 and Figure 8.4). Even the prevention of invasion of an area by this plant will require the support of a number of different groups. Awareness raising should be carried out strategically to engage relevant bodies at the right time. For example, drawing the attention of senior staff in an organisation might be necessary early on in order to gain agreement for funding to set up a management programme; whilst informing the public of a menace on their doorstep might best be left to later on when the management project is better developed and the role, if any, for the public is more clearly defined.

Training and education

In order to achieve maximum impact from the time and money invested in a management project, it is essential that the personnel involved are appropriately trained. Make sure that training needs are identified in relation to the various facets and tasks within the project and develop or look for sources for this training. It can be internal or external to your organisation and can be in the form of written material, for example a manual, a course or gaining work experience, such as work shadowing.

Surveying, turning words into action (control), and restoration

As these aspects are substantial parts of any management plan, they have been given chapters of their own (see Chapters 9, 10 and 11).

Rapid response

Box 8.3 Raising awareness to assist with a management plan (source Child and Wade 2000 Chapter 12, number 4)

Awareness raising does not have to be expensive. A lot can be achieved with:

- A carefully thought out and targeted leaflet (Figure 8.3);
- Well designed and focussed web page;
- An article in a newspaper or magazine (Figure 8.4);
- An information booklet; and
- A workshop on Giant Hogweed and its management.

Planning for an awareness-raising programme should:

- Establish the aims of the awareness raising programme;
- Determine which groups of people need to be aware of the plant and its problems;
- Decide what information will be needed;
- Consider the use of the media;
- Raising funds to cover the cost;
- Consider the implications of the programme; and
- Prepare for the follow up from the programme.

Follow up for the programme could include:

- Answering enquiries and providing advice;
- Collating data;
- Ensuring follow-up support is available, for example, sufficient supplies of protective equipment; or
- Advising colleagues in other sections or agencies, or on neighbouring property of potential implications of the programme.

Where an area that has been recently invaded by Giant Hogweed is identified, a rapid response is essential to quickly eradicate the plant and stop becoming established. Ideally, rapid response should be undertaken to prevent the plants from flowering and setting seed. The procedure for initiating a rapid response should be developed as part of any management scheme so that issues that might cause delay in treatment (for example ownership and responsibility) are resolved beforehand.

Monitoring and surveillance

It is essential to undertake monitoring of the progress and success of a project. Although this is usually implemented well into and at the end of the project, it must be planned for at the outset. It is important to link the way the project's success will be measured to the monitoring output. Typically this will be areas of land or lengths of riverbank cleared of Giant Hogweed by a given date. Well thought through monitoring will provide valuable feedback to inform changes to the management plan and revisions of the policy.

Monitoring forms part of the surveillance used to check for any reinvasion or re-emergence of Giant Hogweed in the project area. Surveillance includes vulnerable areas that have not been invaded yet as well as areas of known invasion and control. Surveillance should include visits to areas identified as being at risk at least annually so that a rapid response can be initiated as soon as invasion occurs.

Giant Hogweed (image from Šárka Jahodová)

9. Surveying

One of the first stages of a Giant Hogweed management programme is to assess the scale of the problem by undertaking a survey. The objectives of the survey should be clearly set out beforehand and the size of the survey area identified. Objectives may range from simply assessing whether Giant Hogweed is present or not, to accurately mapping all of the Giant Hogweed and features relevant to selecting a control method. The size of the survey area could range from a single site, to a catchments scale or national survey (see Chapter 8).

Giant Hogweed surveys should be undertaken by someone confident in the identification of Giant Hogweed. Inexperienced surveyors should meet a minimum training standard, for example reading relevant parts of this book and work shadowing an experienced surveyor on at least one occasion. Training should also be provided to assess site features that are relevant other than just the location of the Giant Hogweed (see below). Where a surveyor struggles to identify plants, photographs can be taken for later identification. These should include close up pictures of leaves, stems and flower or seed heads in addition to overall pictures of the whole plant. Where possible, try to provide a scale for images.

Surveying Giant Hogweed on development / small scale sites

Generally, a technical survey using appropriate instrumentation, for example a Geographical Positioning System unit or equivalent, is most appropriate for Giant Hogweed on a development site. This provides highly accurate mapping, which is needed to select appropriate methods for control and ascertain the extent of the hazard (Figure 9.1). Technical surveys are also easily incorporated into the developer's plans for the site via software such as Auto-Cad or GIS. Where technical instrumentation is not available, the surveyor may use an appropriately scaled site plan to map stands by hand, where possible using measurements taken from site features to triangulate the position of a stand or plant.

Each stand or plant should be given a unique code to distinguish it and key variables recorded for each (Box 9.1). This becomes increasingly important during the control programme as control strategies may differ between stands or plants. Experience shows that once the heavy machinery move onto the site the possibility of seed dispersal increases considerably. It is therefore essential to be able to locate the different areas in relation to the details of the development itself.

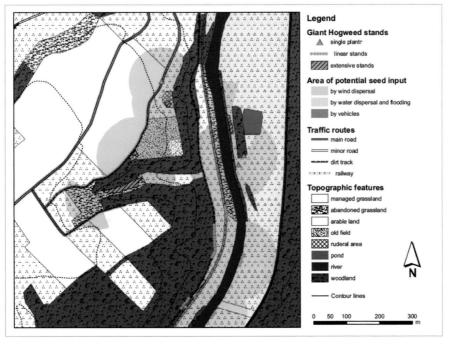

Figure 9.1 GIS map showing the position and potential spread of Giant Hogweed (from Nielsen. et al 2005, Chapter 12, number 19)

Surveying large scale areas

A number of methods can be used for large-scale survey of Giant Hogweed. These can be used in isolation, but are best used in combination to provide the most accurate results.

Desk based surveys

Biological records are well documented in the UK and can be accessed from UK, to regional and local scales. Biological records are rarely up to date but do provide data that have been collected in some cases for over a hundred years. In addition to giving an indication of distribution, they can provide a valuable perspective on the historical spread of the plant.

Biological records for the whole of the UK are kept by the Biological Records Centre (Chapter 12, Useful Links). In addition the National Biodiversity Network (NBN) website hold many of these records as well as other data sources in an easily accessible format (see Chapter 12, Useful Links).

Box 9.1 Survey variables relevant for mapping and monitoring stands of Giant Hogweed (adapted from *Nielsen et al. 2005 Chapter 12, number 19*)

Aim	Parameter	Description
Identification of the stand	Site reference number, date of record and name of recorder	Each site field or plant is assigned a serial number as the stands are identified in the field
Description	Location	Description of the exact location of the plants, map grid reference, GPS coordinates
	Land ownership	Private, public and any other details
	Land register number	Optional
Population	Stage of plant growth	Vegetative, flowering, fruiting, dead
	Area	Area covered by the plant in m^2
	Density	Estimation of the number of plants per m^2
Site Condition	Land use	Agricultural land, river bank, waste ground etc.
	Area access and ground condition	Distance to the nearest paved road and evaluation of ground carrying capacity for heavy machinery
Prediction of impact if not controlled	Plant community	Description of the current plant species under threat due to invasive aliens etc.
	Recreational value	Evaluation in terms of public access to area, proximity to houses and the suitability of the area for recreational purposes
	Risk of erosion of soil, especially into streams or rivers	Density of ground vegetation and the degree of slope with respect to risk of erosion rated as high, medium or low
Management and control	Management history	Status of control, special circumstances to consider
	Control measure	Suggested control method based on an immediate assessment in the field

At a local level, individual areas may have local biological record centres. These usually function at the county scale and are usually associated with a wildlife trust or county council. Data held by the national records centre may not be as up-to-date as those held by the local centres and so these can provide useful additional information for species at this scale.

Aerial surveys

A novel technique that could be used for large-scale Giant Hogweed survey is aerial photography. In full flower during mid-late summer, the huge flower heads can be identified from the air providing a quick and accurate picture of the plant's distribution. Guidance for the use of aerial photography as a survey tool is provided in references in (Chapter 12, number 18).

Field surveys

The only way to obtain complete and current distribution data for Giant Hogweed is to carry out field surveys. Depending on the size of the area this could include a team of surveyors using GPS or relevantly scaled field maps to mark individual plants or groups of plants. The difficulties in identifying Giant Hogweed (Chapter 2) should be considered before conducting the survey. In particular, surveyors should be aware that small plants can very easily be overlooked under the cover of other vegetation. Surveyors should be skilled in identifying young Giant Hogweed plants and surveys should be repeated at regular intervals.

Health and safety is an important concern for field surveyors. They should ensure to take appropriate care when working on their own, near watercourses, on private land and with Giant Hogweed (Box 10.3).

Field surveys can be augmented using support gained from awareness raising campaigns. These could be used to encourage landowners and other stakeholders to send in records, which can be verified by a survey team.

Data storage and representation

The data collected from all of the survey methods could be stored using GIS and used for analysis such as describing spread over time or identification of specifically important corridors of spread. GIS is particularly useful for storing large amounts of data over a large geographic area. If necessary other methods could include a bank of survey maps with associated database. Large scale surveys can make good use of GIS systems to efficiently map Giant

Figure 9.2 Giant Hogweed growing on a river margin

Hogweed locations at a relevant scale onto appropriate base maps. These can be based on Ordnance Survey maps to help locate relevant features and also prepare maps for surveyors to take into the field

Early detection and monitoring

Following control, Giant Hogweed sites should be repeatedly monitored in case of regeneration or new seedling development. This is particularly important, as areas where control has taken place may be particularly sensitive to reinvasion and continuing germination of seeds still present in the seed bank.

Visits should be made at the beginning of the next growing season, usually between March and April. Any new areas of growth should be identified and follow up control undertaken to prevent the possibility of new plants setting seed. Early detection of Giant Hogweed followed by rapid response allows control to be carried out effectively and at low cost (Chapter 8).

Vulnerable habitats that have not yet been invaded should also be monitored (Chapter 8). Particular attention should be paid to:

- Ecologically sensitive sites, for examples Sites of Special Scientific Interest and nature reserves especially where these include rivers (Figure 9.2);

- Habitats where land management has ceased, for example old farmland along rivers;

- Habitats that have been subject to recent disturbance; and

- Sites where Giant Hogweed is well known to be prevalent in the local environs or on a key transport corridor on the site e.g. water course, railway line.

An important facet of monitoring and early detection is awareness by the general public. Targeting people most likely to be in contact with infested areas is an effective way of increasing the chances of locating Giant Hogweed. Groups to target include anglers, walkers, farmers and natural history societies and clubs. Publicity needs to not only include the identification of Giant Hogweed and the health hazards but also provide instruction on how to record sightings. Responsibility for treating Giant Hogweed lies with landowners and they should also be targeted to be made aware of their role.

10. Control methods

Giant Hogweed control should be designed to achieve two key objectives: killing the existing plants and eradicating the remaining seed bank. A variety of methods are available to do both which is most appropriate depends on the resources available and the requirements of the control strategy. Key points to consider before undertaking control are provided in Box 10.1

Cut Giant Hogweed stand, Suffolk

While the adult plants can be killed relatively quickly, either by herbicide application or by cutting the taproot, the seed bank can last for several years (Chapter 4). A control programme should therefore be designed to last as long as the seedbank persists. Remember that once a plant has set seed it will die naturally (Chapter 4). Time and effort are often wasted attempting to control plants that are already programmed for death.

There are three stages of control:

• Prevent any further seed production by removing flower or seed heads as soon as possible taking care in handling the seed heads to capture as many seeds as possible;

• Control the growing plants using one of the methods outlined below and provide continuing control throughout the growing season to eradicate them (March to September);

• Remove the seedbank or provide consistent repeat control to prevent any plants flowering or setting seed until the seed bank is exhausted.

While the stages of control should be the same for any strategy, the way in which these are achieved can vary depending on the situation the plant is in and timescale required for control. Box 10.2 provides a summary of the control methods available for Giant Hogweed and Figure 10.1 provides a decision key to help select the appropriate control option.

Box 10.1 Important points to remember when undertaking control

---

• Ensure health and safety guidance is followed when carrying out Giant Hogweed control and make sure others are also protected (Box 10.3)
• Be aware of legislation, do not spread Giant Hogweed or dispose of it incorrectly (Chapter 6)
• Make sure control encompasses the seedbank as well as adult plants

---

Box 10.2 Control methods

Control measure	Use	Positives	Negatives	Eradication timescale	Cost relative to effort
Seed head removal	Initial reduction of seeds in seed bank	Helps prevent further new growth	Does not control adult plants	Not applicable	Moderate
Taproot cutting	Small stands or stands where sensitive removal is required	Immediately kills adult plants	Does not control the seed bank	Kills adults immediately. Follow up treatment for seedbanks 3-5 for year	High
Handpulling	Very small stands where seedbank is being controlled in sensitive areas	Prevents damage to other species or habitats	Labour intensive. Only suitable for seedlings	Not applicable	High
Mowing / repeat cutting	Large stands where herbicide applications are not appropriate	Avoids herbicide use	Not an immediate kill. Does not control seedbank. Health and safety risk	3-5 years	Moderate
Grazing	Farm land and other areas where livestock can gain long term access	Avoids herbicide use	Livestock must be controlled. Potential for environmental damage	3-5 years	Moderate
Ploughing	Very large stands on non-sensitive sites	Immediate kill of adult plants. Helps to bury seed bank	High level of disturbance. Only appropriate on large sites	Immediate kill. Spot treatment required over 3-5 years	Moderate
Herbicide treatment	Individual plants to large stands	Immediate kill of adult plants	Use restricted in some areas. Does not kill seedbank	3-5 years	Moderate
Excavation and disposal to landfill	Small to medium sized stands on non-sensitive sites	Immediate kill of adults and removal of seedbank	High level of disturbance. Expensive	Immediate	Very high

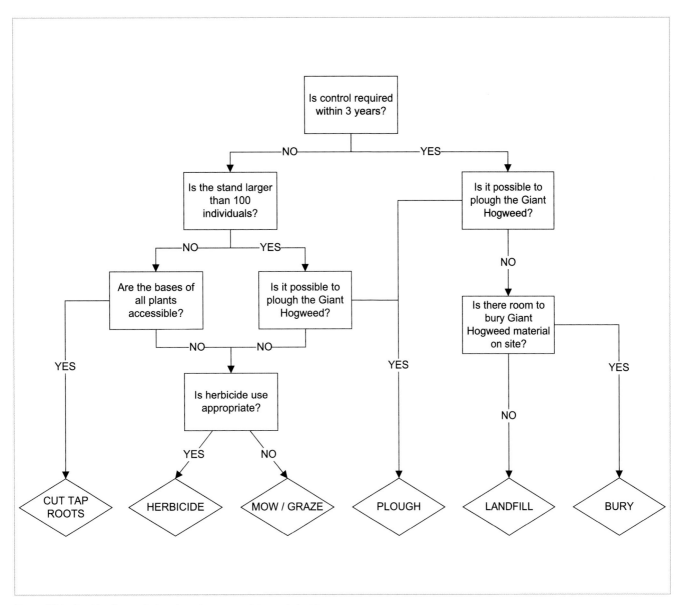

Figure 10.1 Decision key to help select the appropriate control option

Figure 10.2   Bagged seed head

Stage 1 - Removing seed heads

Ideally a control programme should start early in the year, whilst plants are relatively small and have not yet set seed. However, this will not always be possible and in some cases control will necessarily start while plants are in seed. In autumn and winter, old stems may still be present with a large number of seeds remaining on them, although the majority of seeds will drop by the end of December. In these cases, the first aim of the control strategy should be to remove the seed heads in order to prevent or reduce the number of new seeds that are allowed to fall from the parent plant and add to the seedbank.

Seed heads should be individually bagged and then cut; bagging will help prevent the accidental spread of seeds (Figure 10.2). Immature seeds can ripen even after being cut from the parent plant and so all seed heads, no matter how immature, should be disposed of as controlled waste (see Chapter 6).

Physical handling of Giant Hogweed can be dangerous. Safety clothing and equipment should be worn to prevent any chance of the skin coming into contact with the living plant (see Box 10.3). Flower or seed head removal should not be undertaken if stands are so dense as to make access difficult, or if plants are too tall to safely cut the umbels.

Box 10.3 Preventing health hazards when controlling the plant

- Workers attempting to control Giant Hogweed are likely to be at risk of injury from the plant. If attempting to cut Giant Hogweed, full safety protection should be worn including goggles, gloves and full body covering. Synthetic, water-resistant fibres are most suitable as natural fibres such as cotton absorb sap and allow it to come into contact with skin. Strimmers should not be used to cut Giant Hogweed as there is a high risk of inhaling or coming into direct contact with airborne sap.

- Exposed skin must not be allowed to brush against any parts of the plant. Sap-contaminated protective clothing must be removed with care and thoroughly washed or disposed of.

- Individuals can vary in their sensitivity to Giant Hogweed. It is recommended that greater caution be used and exposure to pollen and potentially airborne sap is brief during initial encounters with the plant.

Stage 2 – Eradication of the plants

Herbicide control methods

Herbicides are effective for killing Giant Hogweed (Figure 10.3). Often only one application is enough to kill adult plants, however, re-application throughout the year is necessary to treat new plants that quickly grow to replace the dead ones. Herbicide control will not kill the seeds in the seed bank and therefore applications must be repeated over a number of years to eradicate new plants that grow in following years.

A variety of herbicides available for use in the UK are effective against Giant Hogweed (Box 10.4). The standard chemical available for all situations is glyphosate, a broad-spectrum herbicide, which controls all vegetation, including grasses. The other available chemicals are selective herbicides for use with broad-leaved plants. The careful choice of herbicide for any given situation is important for

Figure 10.3 The before (a) and after effects (b) of herbicide use (images Tim Barratt)

ensuring accidental damage is not done to non-target plants.   Particular care should be taken when applying herbicides near to desirable vegetation such as within amenity land or site of conservation interest. Where Giant Hogweed is situated near water, an application must be made to the appropriate water regulatory body (Environment Agency/Local Rivers Purification Authority or in Scotland the Scottish Environment Protection Agency) for consent to use pesticides in or near water.

The basic stages of any herbicide control programme are:

- Start control in late April to allow the majority of plants to grow to approximately 50cm tall with full leaves;

- Continue to monitor the site and expect to repeat treatment (either blanket or spot spraying) at least once in the same year, preferably three of four times;

- Repeat the process in consecutive years to treat re-growth from the seedbank, which could be up to about four years; and

- Where possible, implement a programme of revegetation (Chapter 11).

Box 10.4 Review of herbicides available for use in the UK that are effective against Giant Hogweed

Herbicide	Persistence in soil (months)	Approved for use near water[3]	When to apply	Selectivity	Human Toxicity[1]	Time taken to achieve control
glyphosate	0	Yes	April – October	Non-selective	Low	3 years repeated application
2,4-D Amine	1	Yes	April – October	Selective for broadleaved species	Low to Moderate	3 years repeated application
triclopyr	1.5	No	April – October	Selective for broadleaved species	Low	3-5 years repeated application
picloram	24	No	All year[2]	Selective for broadleaved species Highly damaging to trees	Moderate	2 –3 years. Glyphosate usually applied after intial treatment

[1]Effects on human health are dependent upon how much of the herbicide is present and the length and frequency of exposure.
[2]Can be applied directly to soil in winter to form a chemical barrier.
[3]It is illegal to use some herbicides near water. If those that are permissible are used, the EA must be informed.

When dealing with a well-established infestation, it is not always necessary to target the complete size range of available plants in the first instance. Spraying large numbers of first year seedlings can be very destructive, particularly where they are mixed in with other vegetation. Since Giant Hogweed takes three or four years to reach maturity, there is an advantage sometimes in leaving the small plants for the time being, to concentrate on the larger plants. When the small plants have grown in the following year they in turn become good targets.

While private individuals may apply store-bought herbicides to plants on their own property following instructions on the label of the purchased product, a herbicide practitioner must be National Proficiency Tests Council certified to apply herbicides commercially. National Proficiency Tests Council training includes understanding the appropriate use of herbicides, the selection of herbicides and method of application, and the regulations governing their use.

There are a wide range of methods available for the application of herbicides (Box 10.5) and the method used will depend on several factors. The practicalities of carrying out herbicide treatment should also be considered before beginning implementing control (Box 10.6).

Figure 10.4 Herbicide being applied using a knapsack spray

Box 10.5 Common methods of herbicide application

- Knapsack Sprayer (Figure 10.4) – this is a commonly used hand-held sprayer. Plastic tanks, with a capacity of over 10 litres, are carried on the operators back. The tanks are pressurised using a pump, which can be hand or battery-operated. The herbicide is applied via a hand-held lance. To minimise the risk of drift the lance can be fitted with a guard or hood;

- Long Lance Sprayer – the application of herbicide using a long lance applicator is useful in areas where accessibility is an issue, such as on riverbanks or steep slopes;

- Controlled Droplet Applicator (CDA) – CDA sprayers make it possible to apply herbicides over larger areas. CDA sprayers increase herbicide efficiency as they are designed for low volume herbicide applicators. CDA sprayers incorporate a spinning disc. These produce a more even droplet size than traditional sprayers, and this reduce the risks of drift and damage to other vegetation. Larger droplets can bounce off larger plants and increase losses to soil and the surrounding area;

- Weedwiping – this facilitates the precise targeting of weeds and removes the risk of contaminating the surrounding vegetation. With this method there is no risk of spray drift and less herbicide is used compared to conventional spraying. However, weedwiping may prove to be problematic once the flower stalks of Giant Hogweed have been produced. This is only really suitable for areas of 1 or 2 plants;

- Boom Spraying – for large stands, herbicide sprays can be delivered from a boom attached to a tractor or other suitable vehicle.

Box 10.6 Points to consider before applying herbicides

- When plants are at full height stands may be virtually impenetrable and spraying the highest leaves may be difficult, dangerous and against the directions for use of the herbicide - in such cases herbicide application should be delayed until the following spring if possible, or, if necessary, cutting and allowing the plants to re-grow, or the use of long lance herbicide application could be considered;

- Giant Hogweed frequently grows along steep riverbanks. The practicalities of traversing the bank should be considered along with other options, for example use of long lance or spraying from a boat;

- Giant Hogweed frequently grows along railway lines and where herbicide operatives are required to access railway land, each will have to be approved and appropriately licensed by the relevant authority. Any persons operating on railway lines must hold a Personal Track Safety license; and

- Spray drift from herbicides may result in the damage of non-target plants and as Giant Hogweed control sometimes requires applying herbicides at height, spray drift may be more likely; regulations governing the application of chemicals must be carefully followed.

The use, storage and disposal of herbicides are regulated by various pieces of UK and European legislation. There are also statutory Codes of Practice, which give advice to users, suppliers and others of their legal responsibilities and how to comply with them. The most important pieces of legislation relevant to the use of herbicides are:

- Control of Substances Hazardous to Health Regulations 1988 (COSHH) covering the use of all pesticides, placing an emphasis on assessing the risks associated with the use of any substance hazardous to health before it is used: whether it is necessary to use the pesticide must also be considered, along with potential risk to humans, animals and the environment; and

- Control of Pesticides Regulations 1986 (COPR) as amended by the Control of Pesticides Regulations 1997, under which only approved products can be sold, supplied, stored, advertised or used. More specifically pesticides can only be used in situations for which their use is currently approved. Lists of approved herbicides are reviewed annually and the lists are available to view on the Pesticides Safety Directorate website (see Chapter 12).

Various bodies are responsible for ensuring compliance with the relevant legislation. The Health and Safety Executive (HSE), local authorities and the agriculture departments share between them the responsibility for enforcement. The Health and Safety Executive is responsible for enforcing controls on the use and storage of herbicides when they are used in relation to work activity by contractors in private dwelling houses, by local authorities and public utility companies. The HSE is also responsible for use in agriculture and the storage of herbicides.

Local authorities enforce controls relating to the advertisement, sale, storage and use for those areas not under the jurisdiction of the HSE. This includes golf courses, sports grounds, gardens, parks and garden centres.

Physical control methods

Cutting the taproot

Giant Hogweed plants can be killed quickly by cutting the main root known as the taproot (Figure 1.2b) that provides it with water, energy and nutrients. The taproot extends deep into the ground immediately below the stem and must be cut at least 15cm below ground level (or the top of the root if slightly buried) to kill it. A sharp spade inserted at an angle into the ground next to the plant is usually sufficient to cut the taproot (Figure 10.5) although mature plants with large taproots can be very difficult to cut and may require several attempts. Once cut, the remains of the taproot should be carefully dug out and disposed of or left to dry out.

Figure 10.5 Cutting taproot

Whilst cutting taproots is an effective way of killing plants, it is highly labour intensive for large stands and does not tackle the seed bank. Smaller plants are also often missed and so control should be repeated later in the growing season. This method is most suited for small stands where precise control is required and long-term management can be provided to control the seed bank.

Repeat cutting of flower heads

Giant Hogweed naturally dies back after flowering and setting seed (Chapter 4). By cutting plants while they are in flower but before they set seed, this natural process can be exploited to control the plants. The plants will complete their life cycle and die off completely, but no seeds will be left to re-grow in the following year.

It is important to note that after cutting Giant Hogweed plants will vigorously re-grow in the same year. This is particularly the case when only the flower heads are removed, as new flowers will quickly re-grow in their place or on secondary stems further down the plant. Intensive monitoring and repeat cutting is essential to prevent new re-growth from flowering and setting seed. If carefully monitored and repeated each year as new plants come into maturity and flower, no new seeds will be produced. Within 3-5 years the seedbank will be exhausted and all of the plants controlled.

The Tweed Forum utilised a similar control method in Scotland and the north of England by endeavouring never to allow Giant Hogweed to flower. A key benefit of this approach was that the objective was clear and easily understood.

Figure 10.6 Handpulling

Figure 10.7 Giant Hogweed growing at the edge of agricultural land

Figure 10.8 Cattle grazing on Giant Hogweed

Handpulling

Handpulling is only effective at removing seedlings and small plants should not be used to control adult plants, as it is ineffective and potentially dangerous. Given the thousands of seeds produced by Giant Hogweed plants this is a highly labour intensive method and is only suitable for removing occasional seedlings (Figure 10.6).

Mowing

The energy of Giant Hogweed is stored in its taproot and year by year this increases until sufficient energy has been accumulated to allow the plant to produce a stem, flower and set seed. By repeatedly mowing the plants and preventing its leaves from fully developing, the energy in the taproot is eventually exhausted and the plant dies. This effect can be seen where Giant Hogweed grows along the boundary of a farmer's field or grass verges, which are regularly cut (Figure 10.7).

Eradicating Giant Hogweed by mowing can take several years (in some cases over seven years) and should only be used where long-term treatment is acceptable. Once cut, Giant Hogweed quickly re-grows and can produce low growing flowers and seed heads and so it is important to monitor the re-growth and provide follow up treatment where necessary (usually at least three times a year).

Methods of mowing, such as flail mowing and strimming, can be dangerous and should be undertaken with extreme care. Full safety clothing should be worn and all machinery carefully cleaned down after use. These methods are more likely to produce air borne sap that can cause respiratory damage and so operatives should wear respiratory protection, works should be undertaken during calm conditions and the general public should be kept away during works.

Grazing

Cattle, sheep, pigs and goats will all eat young Giant Hogweed plants; sheep even grow to favour it over other herbs. Grazing by cattle (Figure 10.8), sheep and goats has a similar effect to mowing; eventually livestock deplete the energy sources of the Giant Hogweed plants and they die. Because the plants roots are left undamaged the plant can persist for many years before eventually being eradicated. However, grazing by pigs can be considerably more effective as they also damage the root system of the plant.

Although grazing is unlikely to be appropriate in many habitats in the UK, for example river banks and urban areas, where feasible it can be an effective solution for control of large Giant Hogweed stands. In some instances livestock used to graze Giant Hogweed have developed skin inflammation and blistering, especially around the lips, face, eyes and underbelly. In general animals with the most pigmentation in their skin (that is darker coloured animals), or those with more hair are less likely to be affected.

Ploughing

Deep ploughing or tilling is a highly effective form of Giant Hogweed control, where it can be applied. By cutting into and turning the soil, the plough both cuts the taproot of the Giant Hogweed and helps to bury the seed bank at a depth that will inhibit growth. This is a method that has mostly been practiced in European countries where there are large agricultural fields of Giant Hogweed that need to be controlled. This could be an equally appropriate method of control in the UK where the plant has invaded the edges of agricultural fields or has formed large dense stands in the open countryside.

Excavation

Where a more immediate solution is required, particularly on development sites, Giant Hogweed can be excavated (Figure 10.9). This provides both an immediate kill by cutting and/or removing the taproot and, if done carefully, can remove the seed bank as well.

In order to remove the Giant Hogweed seed bank, a prediction must be made on the likely distribution of seeds on the site. While it is reasonable to assume that seeds will have fallen at least 10m from parent plants based on wind dispersal figures (see Chapter 4), a more focussed prediction needs to be based on a wide variety of factors including transport along water corridors, roadways, railways and incidental soil movement. A full survey of these factors on site can help to predict the likely distribution and limit the excavation area. The depth of excavation needed is likely to vary, however the seedbank is unlikely to extend deeper than 50cm below surface level; at this depth the taproot of adult plants will also be removed.

The excavation should be undertaken using a mechanical excavator to remove all vegetation as well as soil that could contain seeds of the plant. Remember, Giant Hogweed material, including excavated soil, must not be spread and must be treated in accordance with the Environmental Protection Act (Chapter 6). Extreme care must be used to prevent the spread of material on and off site as the risk of spreading seeds onto neighbouring sites is high. Often

Figure 10.9 Excavating soil contaminated with Giant Hogweed seeds

Figure 10.10 The Lixus weevil

it is appropriate to appoint an Ecological Clerk of Works responsible for overseeing the excavation and ensuring work is undertaken under controlled conditions. Care should also be taken to prevent any injury caused by the plants (see Box 10.3).

Once excavated, Giant Hogweed material must be disposed of appropriately or stockpiled and treated on the site from which it came. Stockpiling is appropriate where Giant Hogweed must be removed from a particular area on the site, but can be moved to another part of the same site for further treatment. In this case the Giant Hogweed material should be stored in an area that can be sectioned off from the rest of the site and allowed to re-grow. It can then be subjected to one of the treatment methods outlined above.

Biocontrol

Considerable research has been undertaken to try to find a biological agent to control Giant Hogweed in Europe. Several possible candidates were identified, including the Lixus weevil (Figure 10.10), however none was found to be suitable. While biocontrol is a valuable control method for many invasive species, research to date has concluded that it is unlikely to be developed as an appropriate tool for Giant Hogweed (for further information on the biological control of Giant Hogweed see Chapter 12, number 30).

Disposal of Giant Hogweed waste

Giant Hogweed material is classified as controlled waste and, if taken off site, must be disposed of responsibly (see Chapter 7). The three most common methods of disposal of this material are incineration, on site burial and disposal to licensed landfill site. Guidance for each method is provided below:

Incineration – Giant Hogweed waste can be incinerated, however this should only be used for the disposal of the taproot or seed heads of Giant Hogweed; the effectiveness of incinerating soil-containing seeds is untested. Thoroughly burning taproots in an open fire on site can be effective; however seeds should be burnt in a more controlled manner. Mobile incinerators can be brought to the site to provide appropriate eradication, however care should be taken during incineration to prevent seeds escaping. Any ash resulting from the burning of Giant Hogweed waste must be retained on the site from which it came. Before undertaking any burning, the relevant authorities including the relevant environmental protection agency (EPA) (Box 7.1) and local fire authority must be consulted.

Burial on site - Giant Hogweed seeds are unlikely to germinate at depth and it is therefore possible to control them by burying them on the site from which they came. Both seed heads and soil containing seeds could be controlled in this way. Burial should be undertaken so that all contaminated material is at least 1m below the surface level of the site. Seeds persist for several years after burial and so the burial site should not be disturbed after disposal. The relevant Environmental Protection Agency (see Box 7.1) should be consulted before undertaking burial on site and the position of the buried material should be clearly mapped and documented for the landowner and regulatory agency.

Disposal to authorised landfill - If taken away from the site of origin, Giant Hogweed must be disposed of at a landfill site authorised to take it. The waste must be transported to the landfill site by a hauler, licensed to transport contaminated waste, and relevant waste transfer documentation must be provided for every load. At all stages due care must be taken to ensure contaminating material is not spilled in transit. The landfill site must be informed a reasonable time before works begin to give them time to prepare to accept it. The charge for disposing of Giant Hogweed waste can vary depending on the landfill site but is likely to be in the region of £30 - £80 per tonne (cost in 2007). A significant proportion of the cost of disposal is due to a levy from the Landfill Tax and an exemption can be granted from this because Giant Hogweed is considered a contaminant and therefore its removal helps in the process of regeneration. In order to receive Landfill Tax exemption, HM Revenul and Customs Excise must be satisfied on a number of points, including that the contamination has ceased, the exact quantity of contaminated material has been calculated and that the removal of the contamination is necessary.

Stage 3 - Eradication of the seedbank

The strategies outlined above control the seedbank over a long period of time using repeated treatment. Only excavation combined with disposal, or ploughing provides a more instant solution. Whichever control strategy is used, sites should be monitored for several years after control for re-growth or re-invasion and follow up treatment provided as necessary (Chapters 8 and 9). Where possible, planned revegetation should be encouraged in order to stabilise areas that have been subjected to control (Chapter 11). This helps to protect the site from reinvasion and restores the ecological function of the controlled area.

Giant Hogweed regrowing among other vegetation after control

11. Revegetation

Bare patches of ground are often left as a result of Giant Hogweed control. This can lead to environmental damage, for example riverbank erosion, as well as making the habitat more susceptible to reinvasion by other invasive species, for example Japanese Knotweed or Winter Heliotrope. However, bare patches also provide an opportunity to reinstate appropriate vegetation and carry out habitat enhancement.

The type of control method used to kill Giant Hogweed can result in differences in the need for and type of revegetation. For example, a considerable proportion of soil would need to be replaced if excavation were used, whereas herbicide application leaves the soil intact but can leave large areas with no vegetation whatsoever. Using persistent herbicides (Box 10.4) may mean that only a carefully selected grass mix could be used for revegetation. Selective cutting will allow some vegetation to remain enabling rapid recolonisation of exposed areas. The state of the ground resulting from control should be considered when choosing a revegetation strategy.

There are three broad options available:

- Take no action and allow the vegetation to establish naturally;

- Revegetate the patches with predetermined plant species, or

- Revegetate the patches with the specific aim of stabilizing the soil surface to prevent erosion.

For small patches of exposed soil, vegetation will establish naturally and probably quickly, therefore it might not be necessary to provide any seeding or planting. Check on what plants colonise the patches and where necessary weed out any undesirable species. If there is a risk of erosion revegetation is essential (see below) as even small patches of bare soil can create a significant problem.

Where the area left bare after treatment is large, for example more than 100m^2, consideration should be given to controlled revegetation. The mix and nature of the vegetation to be grown on exposed areas of soil can be controlled by selecting and sowing appropriate seed mixes, by planting seedlings, or using turf. Establish the aim of revegetation early in the project so that you can choose and order a seed mix, plants or turf in good time. The species mix could achieve a number of aims:

- Re-establishing the diversity of the vegetation lost or declined as result of the Giant Hogweed using a seed mix;

- Creating new habitat, for example planting a woodland strip;

- Reinstating a grassland area, for example the edges of a playing field using turf; and

- Achieving landscaping as part of a larger scheme, for example a housing development.

Whatever the aim of revegetation, consideration should be given to the following factors:

- The vigour of the seeds or plants to be used, to ensure that they are able to establish themselves and able to withstand competition from other plants. If necessary they should be able to withstand regular cutting or mowing, and, or flooding;

- The provenance of the seeds or plants is important, particularly if the aim of the revegetation is to achieve biodiversity enhancement;

- Whether or not a specialist landscape contractor is required to undertake the revegetation;

- The density of sowing and planting bearing in mind cost and the likelihood of the failure of some plants to survive and the potential of undesirable species to re-colonise;

- The time of year for sowing and planting;

- Preparing the ground prior to sowing or planting; and

- Aftercare, such as weeding and replacement planting. For example using a selective herbicide suitable for broadleaved weeds (including newly emerged seedlings of Giant Hogweed) in a developing grass sward (see Chapter 10 and check what herbicides are permitted for use).

Revegetation after Giant Hogweed control is similar to revegetation in other situations and good advice is available (Chapter 12, numbers 10, 26 and 41).

A particular aim of revegetation could be to stabilize the exposed soil to prevent erosion. This is especially relevant along rivers and in floodplains that are subject to flooding and the erosive action of water. Erosive action can lead to the collapse of banks and the undermining of bank side trees, as well as exacerbating the risk of flooding downstream. A well-planned revegetation programme can reduce these risks. Where Giant Hogweed has been controlled using herbicide its deep taproot dies away leaving substantial holes. On a riverbank these can act as focal points for erosion, especially when a bank side is inundated, exacerbating the effect of the erosion with potentially dramatic and serious results.

Where selective control methods are used, such as taproot cutting or application of selective herbicides, revegetation can begin while a control programme is being undertaken. For example, if the plan is to plant trees, these can go in early in the programme. Likewise, if a geotextile is to be used in conjunction with seeding, this could be useful in suppressing re-growth of Giant Hogweed and laid down early in the year (see below). On the other hand, revegetation that is begun too early could be out-competed by re-growth of Giant Hogweed if the latter remained a dominant component of the vegetation.

The most commonly used means of revegetation is the sowing of a seed mix that comprises mainly or wholly grass species (Box 11.1). The factors listed above should be considered in relation to the specific aim of the revegetation. On river banks and flood plains those native species that are already locally abundant are likely to be good plants to use for re-establishment. If the risk of erosion is high, for example a steep sided riverbank that is inundated annually; consideration should be given to using a geotextile such as biodegradable

---

Box 11.1 Re-establishment of grasslands in a National Park after the control of Giant Hogweed (*Heracleum sosnowskyi*), Latvia

---

Large areas of the Gauja National Park, Latvia were infested by dense stands of Giant Hogweed as a result of its cultivation as a fodder crop. The Giant Hogweed was controlled using herbicide treatment leaving large areas of denuded ground. After soil cultivation, for example harrowing, grass mixtures were sown at high seed densities (4,000 emerging seedlings/m^2). Native grass species and cultivars were chosen that were known to be competitive, able to produce dense swards, and made good growth after repeated cutting. Examples of grass mixtures that proved to be suitable were: Cock's-foot (*Dactylis glomerata*), and Red Fescue (*Festuca rubra*) (50:50) and Italian Rye-grass (*Lolium perenne*), Cock's-foot and Smooth Meadowe-grass (*Poa pratensis*) (12:35:53). A selective herbicide suitable for controlling broadleaved weeds in the developing grass sward (including newly emerged seedlings of Giant Hogweed) was used as a single application during the vegetative period. Frequent cutting of the re-established grass sward was undertaken when the height of the Giant Hogweed seedlings reached 20-30cm. (from Nielsen *et al* 2005, Chapter 12, number 19)

---

matting. This would be secured to the soil surface and sowed with a grass mix to allow grasses to grow through it. In time, the geotextile would decay leaving a well-established grass sward to protect the riverbank.

In some case, the establishment of woodland or planting trees is appropriate. Where the opportunity arises to plant woodland along a river corridor or in a floodplain, consideration should be given to creating wet woodland, a specific type of woodland under threat in the United Kingdom (Chapter 12, number 34).

Management of the newly established vegetation is usually essential. For example, mowing a grassed area will encourage vigorous root growth, necessary for the stabilisation of the soil and also useful in suppressing any re-growth of Giant Hogweed. Weed control using either hand-pulling or a herbicide may be necessary to suppress weeds that occur in the new vegetation, especially other invasive non-native species such as Himalayan Balsam, Winter Heliotrope and Japanese Knotweed in addition to the native undesirable species such as Broad-leaved Dock and Common Nettle. The new vegetation might not establish uniformly across the revegetated area, for example due to patches of flooding and erosion. Replacement sowing or planting would be necessary to make good such bare patches and losses, for example of trees that have died or have otherwise been lost.

Butterfly resting on Giant Hogweed leaf

12. Further reading

References

1. **Caffrey, J.M. (1994)** Spread and management of *Heracleum mantegazzianum* (Giant Hogweed) along Irish river corridors. In: de Waal, L.C., Child, L.E., Wade, P.M. and Brock, J.H. (eds), Ecology and Management of Invasive Riverside Plants. John Wiley & Sons Ltd, Chichester. pp. 76-76.

2. **Caffrey, J.M. (1999)** Phenology and long-term control of Heracleum mantegazzianum. Hydrobiologia 415: 223-228.

3. **Child, L.E. and de Waal, L.C. (1997)** The use of GIS in the management of Fallopia japonica in the urban environment. In: Brock, J.H., Wade, M., Pysek, P. and Green, D. (eds), Plant Invasions: Studies from North America and Europe. Backhuys Publishers Leiden, : 207-220.

4. **Child, L.E. and Wade, P.M. (2000).** The Japanese Knotweed Manual. Packard Publishing Limited, Chichester.

5. **Child, L.E., Wade, P.M. and Brock, J.H. (eds).** Ecology and Management of Invasive Riverside Plants. John Wiley & Sons Ltd, Chichester.

6. **Clegg, L. M. and Grace, J. (1974)** The distribution of Heracleum mantegazzianum. Transactions of the Botanical Society of Edinburgh 42: 223-229.

7. **Dodd, F.S., de Waal, L.C., Wade, P.M. and Tiley, G.E.D. (1994)** Control and management of Heracleum mantegazzianum (Giant Hogweed). In: de Waal, L.C., Child L.E., Wade, P.M. and Brock, J.H. (eds). Ecology and Management of Invasive Riverside Plants. John Wiley & Sons Ltd, Chichester. pp. 111-126.

8. **Environment Agency (2003)** Guidance for the control of invasive weeds in or near fresh water. Environment Agency, Bristol.

9. **Environment Agency (2006)** The Knotweed Code of Practice: Managing Japanese Knotweed on Development Sites. Environment Agency, Bristol.

10. **Gilbert, O.L. and Anderson, P. (1998)** Habitat Creation and Repair. Oxford University Press.

11. **Hakansson, S. (2003)** Weeds and Weed Management on Arable Land: An Ecological Approach. CABI Publishing, Walling ford UK.

12. **Hemphill, R.W. and Bramley, M. (1990)** Protection of River and Canal Banks. CIRIA, Butterworths, London.

13. **Jahodova, S., Trybush, S., Pyšek, P., Wade, M. and Karp, A. (2007)** Invasive species of Heracleum in Europe: an insight into genetic relationships and invasion history. Diversity and Distributions 13: 99-114 (http://www.giant-alien.dk/pdf/Jahodova_et_al_2007.pdf),

14. **Krinke, L., Moravcová, L., Pyšek, P., Jarošík, V., Pergl, J. and Perglová, I. (2005)** Seed bank in an invasive alien Heracleum mantegazzianum and its seasonal dynamics. Seed Science Research 15: 239-248. (http://www.giant-alien.dk/pdf/Krinke_et_al_SeedSciRes2005.pdf).

15. **Lindstrom, H. and Darby, E. (1994)** The Heracleum mantegazzianum (Giant Hogweed) problem in Sweden: Suggestions for its management and control. In: de Waal, L.C., Child L.E., Wade, P.M. and Brock, J.H. (eds). Ecology and Management of Invasive Riverside Plants. John Wiley & Sons, Chichester. pp. 93-100.

16. **Moracova, L., Pyšek, P., Pergl, J., Perglová, I. and Jarošík, V. (2006)** Seasonal pattern of germination and seed longevity in the invasive species Heracleum mantegazzianum. Preslia 78: 287-301 (http://www.giant-alien.dk/pdf/Moravcova_Pysek_et_al_Preslia_2006.pdf).

17. **Moravcová, L., Perglová, I., Pyšek, P., Jarošík, V. and Pergl, J. (2005)** Effects of fruit position on fruit mass and seed germination in the alien species Heracleum mantegazzianum (Apiaceae) and the implications for its invasion. Acta Oecologica 28: 1-10 (http://www.giant-alien.dk/pdf/Moracova_et_al_ActaOecol2005.pdf).

18. **Müllerová, J., Pyšek, P., Jarošík, V. and Pergl, J. (2005)** Aerial photographs as a tool for assessing the history of invasion by Heracleum mantegazzianum. Journal of Applied Ecology 42: 1042-1053 (http://www.giant-alien.dk/pdf/Mullerova_et_al_2005.pdf).

19. **Nielsen, C., Ravn, H.P., Nentwig, W. and Wade, P.M. (eds) (2005)** The Giant Hogweed Best Practice Manual. Guidelines for the management and control of an invasive weed in Europe. Forest & Landscape Denmark, Hørsholm.

20. **Ochsmann, J. (1992)** Riesen-Bärenklau (Heracleum sp.) in Deutschland, Morphologie, Ökologie, Verbreitung und systematische Einordnung, in Systematisch-Geobotanischen Institut., Georg- August- Universität: Göttingen.

21. **Ochsmann, J. (1996)** Heracleum mantegazzianum Sommier et Levier (Apiaceae) in Deutschland – Untersuchungen zur biologie, verbreitung, morphologie und taxonomie. Feddes Repertorium 107: 557-595 (in German).

22. **Otte, A. and Franke, R. (1988)** The ecology of the Caucasian herbaceous perennial Heracleum mantegazzianum Somm, et Lev. (Giant Hogweed) in cultural ecosystems of Central Europe. Phytocoenologia 28: 205-232.

23. **Pergl, J., Perglová, I. and Pyšek, P. (2005)** Age structure of Heracleum mantegazzianum populations studied by using herbchronology [In Czech, summary in English]. Zprávy Ceske Botanicke Spolecnosti, Praha, 40, Mater. 20: 121-126 (http://www.giant-alien.dk/pdf/Pergl_et_al_2005.pdf).

24. **Pergl, J., Perglová, I., Pyšek, P. and Dietz, H. (2006)** Population age structure and reproductive behavior of the monocarpic perennial Heracleum mantegazzianum (Apiaceae) in its native and invaded distribution ranges. American Journal of Botany 93: 1018-1028 (http://www.giant-alien.dk/pdf/Pergl_et_al_AmJBot 2006.pdf).

25. **Perglova, I., Pergl, J. and Pyšek, P. (2006)** Flowering phenology and reproductive effort of the invasive alien plant Heracleum mantegazzianum. Preslia 78: 265-285 (http://www.giant-alien.dk/pdf/Perglova_Pergl_Pysek_Preslia_2006.pdf).

26. **Perrow, M.R. and Davy, A.J. (2002)** Handbook of Ecological Restoration. Volumes 1 and 2.. Cambridge University Press, Cambridge.

27. **Pysek, P. (1991)** Heracleum mantegazzianum in the Czech Republic: the dynamics of spreading from the historical perspective. Foilia Geobotanica et Phytotaxonomica 26: 439-454.

28. **Pysek, P. and Pysek, A. (1995)** Invasion by Heracleum mantegazzianum in different habitats in the Czech Republic. Journal of Vegetation Science 6: 711-718.

29. **Pysek, P. and Richardson, D.M. (2007)** Traits associated with invasiveness in alien plants: where do we stand? In: Nentwig, W. (ed) Biological Invasions. Springer, Berlin, pp. 97-125.

30. **Pysek, P., Cock, M.J.W., Nentwig, W. and Ravn, H. P. (2007)** Ecology and Management of Giant Hogweed (Heracleum mantegazzianum). CABI, Wallingford, UK.

31. **Pysek, P., Kopekcy, M., Jarosik, V. and Kotkova, P. (1998)** The role of human density and climate in the spread of Heracleum mantegazzianum in the Central European landscape. Diveristy and Distributions 4: 9-16.

32. **Pysek, P., Krinkle, L., Jarosik, V., Perglova, I., Pergl, J. and Moravcova, L. (2007)** Timing and extent of tissue removal affect reproduction characteristics of an invasive species Heracleum mantegazzianum. Biological Invasions (in press

doi 10.107/s10530-006-90380).

33.   **Pysek, P., Mandak, B., Francirkova, T. and Prach, K. (2001)** Persistence of stout clonal herbs as invaders in the landscape: a field test of historical records. In: Brundu, G., Brock, J., Camarda, I., Child, L. and Wade, M (eds) Plant Invasions: Species Ecology and Ecosystems Management. Backhuys, Leiden, pp. 235-244.

34.   **Rodwell, J.S. and Patterson, G. (1994)** Creating New Native Woods: 8, Wet Woodlands. Forestry Authority Bulletin 112. Forestry Authority, Edinburgh.

35.   **Sheppard, A.W., Shaw, R.H. and Sforza, R. (2006)** Top 20 environmental weeds for classical biological control in Europe: a review of opportunities, regulations and other barriers to adoption. Weed Research 46: 93-117.

36.   **Thiele, J. and Otte, A. (2006)** Analysis of habitats and communities invaded by Heracleum mantegazzianum Somm. et Lev. (Giant Hogweed) in Germany. Phytocoenologia 36: 281-320 (http://www.giant-alien.dk/pdf/Thiele_and_Otte_2006.pdf).

37.   **Tiley, G.E.D. and Philp, B. (1994)** Heracleum mantegazzianum (Giant Hogweed) and its control in Scotland. In: de Waal, L.C., Child, L.E., Wade, P.M. and Brock, J.H. (eds), Ecology and Management of Invasive Riverside Plants. John Wiley & Sons, Chichester. pp. 101-109.

38.   **Tiley, G.E.D., Dodd, F.S. and Wade, P.M. (1996)** Biological flora of the British Isles: 190. Heracleum mantegazzianum Sommier et Levier. Journal of Ecology 84: 297-319.

39.   **Vivash, R. and Janes, M. (2002)** Manual of River Restoration Techniques – 2002 update. The River Restoration Centre, Arca, Bedford.

40.   **Wade, P.M., Darby, E.J., Courtney, A.D. and Caffrey, J.M. (1997)** Heracleum mantegazzianum: a problem for river managers in the Republic of Ireland and the United Kingdom. In: Brock, J.H., Wade, P.M., Pysek, P. and Green, D. (eds) Plant Invasions: Studies from North America and Europe. Backhuys, Leiden, The Netherlands, pp. 139-151.

41.   **Ward, D., Holmes, N.T.G. and José, P. (1994)** The New Rivers and Wildlife Handbook. RSPB, National Rivers Authority and the Wildlife Trusts, Sandy.

42.   **Weber, E. (2003)** Invasive Plant Species of the World: A Reference Guide to Environmental Weeds. CABI Publishing, Wallingford, UK.

43.   **Williamson, M. (1998)** Measuring the impact of plant invaders in Britain. In: Starfinger, U., Edwards, K., Kowarik, I. and Williamson, M. (eds) Plant Invasions: Ecological Mechanisms and Human Responses. Backhuys, Leiden, The Netherlands, pp. 57-68.

44.   **Williamson, M., Pysek, P., Jarosik, V. and Prach, K. (2005)** On the rates and patterns of spread of alien plants in the Czech Republic, Britain and Ireland. Ecoscience 12: 424-433.

45.   **Williamson, J.A. and Forbes J.C. (1982)** Giant Hogweed (Heracleum mantegazzianum): Its spread and control with glyphosate in amenity areas. Proceedings of the 1982 British Crop Protection Conference – Weeds: 967-972.

46.   **Wittenberg, R. (2004)** The beauty and the beast - two faces of a plant. Aliens 18: 9.

47.   **Wittenberg, R. and Cock M.J.W. (2001)** Invasive Alien Species: a Toolkit of Best Prevention and Management Practices. CABI Publishing, Wallingford, UK.

Useful links

Biological Records Centre
http://www.brc.ac.uk/

Biopix
http://www.biopix.dk

British Water Ways
http://www.britishwaterways.co.uk/

Cornwall Knotweed Forum
http://www.projects.ex.ac.uk/knotweed/

Countryside Council for Wales
http://www.ccw.gov.uk/

Defra
http://www.defra.gov.uk/

Defra Horticulture Code of Practice
http://www.defra.gov.uk/wildlife-countryside/non-native/pdf/non-nativecop.pdf

**Defra Review of Non-native Species Policy:
report of the working group**
http://www.defra.gov.uk/wildlife-countryside/resprog/findings/non-native/index.htm

Environment Agency
http://www.environment-agency.gov.uk/

Environment Agency Wales
http://www.environment-agency.gov.uk/regions/wales/

Environment and Heritage Service (Northern Ireland)
http://www.ehsni.gov.uk/

Giant Hogweed Best Pracice Manual
http://www.giant-alien.dk/

Inland Waterways Association
http://www.waterways.org.uk/Home

Medway Valley Countryside Partnership
http://www.medwayvalley.org/

National Biodiversity Network
http://www.nbn.org.uk/

Natural England
http://www.naturalengland.org.uk/

Network Rail
http://www.networkrail.co.uk/

Pesticides Safety Directorate
http://www.pesticides.gov.uk/

Scottish Environmental Protection Agency
http://www.sepa.org.uk/

Scottish Executive Environment and Rural Affairs Department
http://www.scotland.gov.uk/About/Departments/ERAD

Scotish National Heritage
http://www.snh.org.uk/

Tamar Valley Invasives
http://www.tamarvalley.org.uk/

Tweed Forum
http://www.tweedforum.com

Cattle grazing on Giant Hogweed in the Caucasus Mountains

Glossary

active ingredient   The component of the pesticide which kills or controls the target pest.

adjuvant   A substance or substances which when added to a pesticide increases the efficacy of the treatment.

alien   See naturalised - or A plant whose native range lies outside the British Isles.

alluvial   Of or relating to a fine-grained fertile soil consisting of mud, silt and sand deposited by flowing water, that is alluvium.

annual   A plant that completes its life-cycle from germination as a seed to production of seed followed by death within a single year.

biocontrol   See biological control.

biological control   Artificial control of pests and diseases by using other organisms, for example introducing a fungal disease of Giant Hogweed in the Caucasus into Western Europe in order to control the plant there.

bract   Whorl of small leaves at base of main umbel.

British Isles   The geographical area including the United Kingdom, the Isle of Man and the Republic of Ireland.

catchment management plan   The plan developed for the management of the land use and/or river or stream/creek within the catchment of that watercourse.

controlled waste   Household, industrial, commercial and special waste.

database   A body of information stored, usually in a computer, from which particular pieces of information can be retrieved when required.

dormancy   A resting condition in which the growth of an organism is halted and metabolic rate is slowed down.

ecosystem   A unit consisting of a community of organisms and the environment in which the live and also interact, for example a wood.

exponential growth   occurs when the growth rate of a population is proportional to the size of the population.

family   A group of similar genera of taxonomic rank below order and above genus; with plants the, the names usually end in –aceae, for example Apiaceae.

fecundity   The reproductive capacity of an organism. For example, Giant Hogweed can reproduce quickly, therefore it has high fecundity.

germination   The beginning of the growth from a seed, usually following a period of dormancy.

genus   A taxonomic group lower than a family consisting of closely related species, for example Heracleum, (plural = genera).

genera	See genus.
Geographical Information System	(abbreviated to GIS) A powerful set of tools for collecting, storing, retrieving at will, transforming and displaying data from the real world
herbaceous	Usually refers to tall herbs that die down during the winter and survive as underground or perennating parts until conditions improve.
herbicide	A chemical used to kill weeds.
hybrid	A plant originating after the fertilization of one species by another species.
inflorescence	A flowering shoot or an aggregation of flowers.
introduced	In this context, refers to the process of either knowingly or unknowingly bringing a plant into an area outside its native range.
invertebrate	A collective name for all those animals which do not possess a backbone or vertebra such as insects and spiders.
journal	A publication, usually published quarterly, of papers or articles of a learned nature.
lowland	Relatively low ground or a low generally flat region.
mericarp	Single seeded portion of dried fruit.
monoculture	The cultivation of a single crop.
naturalised	An introduced species that is permanently established
naturalization	Process of species adapting to places where they are not indigenous.
native	Belonging naturally to a region, for example Giant Hogweed is native to the Caucasus.
nutrient	A nourishing substance, in this context with respect to plant growth.
ornamental	A plant grown for ornament or beauty.
perennation	Survival from year to year by vegetative means. The vegetative parts of the plant, which survive the inclement period, are known as the perrenating parts, for example the rhizome of Japanese knotweed.
perennial	A plant that continues its growth from year to year.
photodermatitis	A reaction of the skin to the UV rays of the sun.
photosynthesis	The use of energy from light to drive chemical reactions in plants.
phytophagous	Plant-eating.
pollinator	An organism which effects the transference of pollen from anther to stigma, usually insects.
predation	Relationship between two species of animal in a community in which one hunts, kills and eats the other.
press release	An official report or statement supplied to the press.

propagule	Any part of a plant capable of growing into a new organism, for example spore, seed or cutting.
riparian	Of or inhabiting a riverbank.
risk assessment	the process whereby the chances of a particular non-native species causing problems after introduction to a country are assessed, based upon previous knowledge of the behaviour of the species and its relatives in its native
ruderal	Applied to plants which inhabit old fields, waysides or waste land.
section	A taxonomic group which is the sub-division of a genus.
self-fertilisation	Fertilisation by the union of the male and female reproductive cells from the same individual.
seminatural	That which is partly modified by man's influence.
sepal	One of the parts of a calyx usually green and leaf-like.
Site of Special Scientific Interest	An area identified for its valuable wildlife habitat or geological environment, which is protected under UK legislation.
species	A taxonomic group of individuals having common characteristics and placed under a genus.
stamen	The pollen producing part of a flower consisting of the anther and the filament.
stand	A growth or a crop of plants.
stigma	That part of the plant, the surface of the carpel, which receives the pollen.
style	A stalk-like structure of the carpel bearing the stigma.
sulphur dioxide	A suffocating gas discharged into the atmosphere as a part of volcanic processes and in waste from industrial processes.
tap root	Primary root of the plant that grows vertically downwards.
taxa	See taxon.
taxon	A biological category, a taxonomic group or unit, for example, family, genus, species and variety.
taxonomy	The classification of plants and animals.
translocate	The process by which materials for example, nutrients are transported within a plant.
umbel	A flat-topped flower cluster where all flower stalks radiate from the same point.
United Kingdom	Since 1922, the official title for the kingdom consisting of England and Wales, Scotland and Northern Ireland (abbreviated to UK).
understorey	Collectively the trees in a forest below the upper canopy cover.
viability	Being capable of germinating, living and surviving.

Index

Giant Hogweed growing near mile post on the Scottish Borders